Welding
Skills, Processes and Practices for Entry-Level Welders

Book 2

- Gas Metal Arc Welding
- Flux Cored Arc Welding

First Edition

Larry Jeffus

Lawrence Bower

DELMAR
CENGAGE Learning

Australia • Brazil • Japan • Korea • Mexico • Singapore • Spain • United Kingdom • United States

Welding: Skills, Processes and Practices for Entry-Level Welders: Book Two
Larry Jeffus/Lawrence Bower

Vice President, Editorial: Dave Garza

Director of Learning Solutions:
Sandy Clark

Executive Editor: David Boelio

Managing Editor: Larry Main

Senior Product Manager: Sharon
Chambliss

Editorial Assistant: Lauren Stone

Vice President, Marketing: Jennifer
McAvey

Executive Marketing Manger:
Deborah S. Yarnell

Senior Marketing Manager:
Jimmy Stephens

Marketing Specialist: Mark Pierro

Production Director: Wendy Troeger

Production Manager: Mark Bernard

Content Project Manager: Cheri Plasse

Art Director: Benj Gleeksman

Technology Project Manager:
Christopher Catalina

Production Technology Analyst:
Thomas Stover

For product information and technology assistance, contact us at
Professional & Career Group Customer Support,
1-800-648-7450

For permission to use material from this text or product, submit a request online at **cengage.com/permissions.**
Further permissions questions can be e-mailed to
permissionrequest@cengage.com.

Library of Congress Control Number: 2008911231

ISBN-13: 978-1-4354-2790-7
ISBN-10: 1-4354-2790-4

Delmar
5 Maxwell Drive
Clifton Park, NY, 12065-2919
USA

Cengage Learning products are represented in Canada by Nelson Education, Ltd.

For your lifelong learning solutions, visit **delmar.cengage.com**
Visit our corporate website at **cengage.com.**

Notice to the Reader
Publisher does not warrant or guarantee any of the products described herein or perform any independent analysis in connection with any of the product information contained herein. Publisher does not assume, and expressly disclaims, any obligation to obtain and include information other than that provided to it by the manufacturer. The reader is expressly warned to consider and adopt all safety precautions that might be indicated by the activities described herein and to avoid all potential hazards. By following the instructions contained herein, the reader willingly assumes all risks in connection with such instructions. The publisher makes no representations or warranties of any kind, including but not limited to, the warranties of fitness for particular purpose or merchantability, nor are any such representations implied with respect to the material set forth herein, and the publisher takes no responsibility with respect to such material. The publisher shall not be liable for any special, consequential, or exemplary damages resulting, in whole or part, from the readers' use of, or reliance upon, this material.

Printed in the United States of America
2 3 4 5 XX 11 10 09

Brief Contents

Contents

Preface

ABOUT THE SERIES

Welding: Skills, Processes and Practices for Entry-Level Welders is an exciting new series that has been designed specifically to support the American Welding Society's (AWS) SENSE EG2.0 training guidelines. Offered in three volumes, these books are carefully crafted learning tools consisting of theory-based texts that are accompanied by companion lab manuals, and extensive instructor support materials. With a logical organization that closely follows the modular structure of the AWS guidelines, the series will guide readers through the process of acquiring and practicing welding knowledge and skills. For schools already in the SENSE program, for those planning to join, or for schools interested in obtaining certifiable outcomes based on nationally recognized industry standards in order to comply with the latest Carl D. Perkins Career and Technical Education requirements, *Welding: Skills, Processes and Practices for Entry-Level Welders* offers a turnkey solution of high quality teaching and learning aids.

Career and technical education instructors at the high school level are often called upon to be multi-disciplinary educators, teaching welding as only one of as many as five technical disciplines in any given semester. The *Welding: Skills, Processes and Practices for Entry-Level Welders* package provides these educators with a process-based, structured approach and the tools they need to be prepared to deliver high level training on processes and materials with which they may have limited familiarity or experience. Student learning, satisfaction and retention are the target of the logically planned practices, supplements and full color textbook illustrations. While the AWS standards for entry level welders are covered, students are also introduced to the latest in high technology welding equipment such as pulsed gas metal arc welding (GMAW-P). Career pathways and career clusters may be enhanced by the relevant mathematics applied to real world activities as well as oral and written communication skills linked to student interaction and reporting.

Book 1, the core volume, introduces students to the welding concepts covered in AWS SENSE Modules 1, 2, 3, 8 and 9 (Occupational Orientation, Safety and Health of Welders, Drawing and Welding Symbol Interpretation, Thermal Cutting, and Weld Inspection Testing and Codes). Book 1 contains all the material needed for a SENSE program that prepares students for qualification in Thermal Cutting processes. The optional Books 2 and 3 cover other important welding processes and are grouped in logical combinations. Book 2 corresponds to AWS SENSE Modules 5 and 6 (GMAW, FCAW), and Book 3 corresponds to AWS SENSE Modules 4 and 7 (SMAW, GTAW).

The texts feature hundreds of four-color figures, diagrams and tight shots of actual welds to speed beginners to an understanding of the most widely used welding processes.

FEATURES

- Produced in close collaboration with experienced instructors from established SENSE programs to maximize the alignment of the content with SENSE guidelines and to ensure 100% coverage of Level I-Entry Welder Key Indicators.
- Chapter introductions contain general performance objectives, key terms used, and the AWS SENSE EG2.0 Key Indicators addressed in the chapter.
- Coverage of Key Indicators is indicated in the margin by a torch symbol and a numerical reference.
- Contains scores of fully illustrated Practices, which are guided exercises designed to help students master processes and materials. Where applicable, the Practices reproduce and reference actual AWS technical drawings in order to help students create acceptable workmanship samples.
- Each section introduces students to the materials, equipment, setup procedures and critical safety information they need in order to weld successfully.
- Hundreds of four-color figures, diagrams and tight shots of actual welds to speed beginners to an understanding of the most widely used welding processes.
- End of chapter review questions develop critical thinking skills and help students to understand "why" as well as "how."

SUPPLEMENTS

Each book in the Welding Skills series is accompanied by a *Lab Manual* that has been designed to provide hands-on practice and reinforce the student's understanding of the concepts presented in the text. Each chapter contains practice exercises to reinforce the primary objectives of the lesson, including creation of workmanship samples (where applicable), and a quiz to test knowledge of the material. Artwork and safety precautions are included throughout the manuals.

Instructor Resources (on CD-ROM), designed to support Books 1–3 and the accompanying Lab Manuals, provide a wealth of time-saving tools, including:

- An Instructor's Guide with answers to end-of-chapter Review Questions in the texts and Lab Manual quizzes.
- Modifiable model Lesson Plans that aid in the design of a course of study that meets local or state standards and also maps to the SENSE guidelines.
- An extensive ExamView Computerized Test Bank that offers assessments in true/false, multiple choice, sentence completion and short answer formats. Test questions have been designed to expose students to the types of questions they'll encounter on the SENSE Level 1 Exams.
- PowerPoint Presentations with selected illustrations that provide a springboard for lectures and reinforce skills and processes covered in the texts. The PowerPoint Presentations can be modified or expanded as instructors desire, and can be augmented with additional illustrations from the Image Library.
- The Image Library contains nearly all (well over 1000!) photographs and line art from the texts, most in four-color.
- A SENSE Correlation Chart that shows the close alignment of the *Welding* series to the SENSE Entry Level 1 training guidelines. Each Key Indicator within each SENSE Module is mapped to the relevant text and lab manual page or pages.

TITLES IN THE SERIES

Welding: Skills, Processes and Practices for Entry-Level Welders: Book 1, Occupational Orientation, Safety and Health of Welders, Drawing and Welding Symbol Interpretation, Thermal Cutting, Weld Inspection Testing and Codes
(Order #: 1-4354-2788-2)
Lab Manual, Book One (Order #: 1-4354-2789-0)

Welding: Skills, Processes and Practices for Entry-Level Welders: Book 2, Gas Metal Arc Welding, Flux Cored Arc Welding (Order #:1-4354-2790-4)
Lab Manual, Book Two (Order #: 1-4354-2795-5)

Welding: Skills, Processes and Practices for Entry-Level Welders: Book 3, Shielded Metal Arc Welding, Gas Tungsten Arc Welding (Order #:1-4354-2796-3)
Lab Manual, Book Three (Order #: 1-4354-2797-1)

AWS Acknowledgment

The Authors and Publisher gratefully acknowledge the support provided by the American Welding Society in the development and publication of this textbook series. "American Welding Society," the AWS logo and the SENSE logo are the trade and service marks of the American Welding Society and are used with permission.

For more information on the American Welding Society and the SENSE program, visit **http://www.aws.org/education/sense/** or contact AWS at (800) 443-9353 ext. 455 or by email: **education@aws.org**.

Acknowledgments

The authors and publisher would like to thank the following individuals for their contributions to this series:

Garey Bish, *Gwinnett Technical College, Lawrenceville, GA*
Julius Blair, *Greenup County Area Technology Center, Greenup, KY*
Rick Brandon, *Pemiscot County Career & Technical Center, Hayti, MO*
Stephen Brandow, *University of Alaska Southeast, Ketchikan, Ketchikan, AK*
Francis X Brieden, *Career Technology Center of Lackawanna County, Scranton, PA*
John Cavenaugh, *Community College of Southern Nevada, Las Vegas, NV*
Clay Corey, *Washington-Saratoga BOCES, Fort Edward, NY*
Keith Cusey, *Institute for Construction Education, Decatur, IL*
Craig Donnell, *Whitmer Career Technology Center, Toledo, OH*
Steve Farnsworth, *Iowa Lakes Community College, Emmetsburg, IA*
Ed Harrell, *Traviss Career Center, Lakeland, FL*
Robert Hoting, *Northeast Iowa Community College, Sheldon, IA*
Steve Kistler, *Moberly Area Technical Center, Moberly, MO*
David Lynn, *Lebanon Technology & Career Center, Lebanon, MO*
Frank Miller, *Gadsden State Community College, Gadsden, AL*
Chris Overfelt, *Arnold R Burton Tech Center, Salem, VA*
Kenric Sorenson, *Western Technical College, LaCrosse, WI*
Pete Stracener, *South Plains College, Levelland, TX*
Bill Troutman, *Akron Public Schools, Akron, OH*
Norman Verbeck, *Columbia/Montour AVTS, Bloomsburg, PA*

About The Authors

Larry Jeffus is a dedicated teacher and author with over twenty years experience in the classroom and several Delmar Cengage Learning welding publications to his credit. He has been nominated by several colleges for the Innovator of the Year award for setting up nontraditional technical training programs. He was also selected as the Outstanding Post-Secondary Technical Educator in the State of Texas by the Texas Technical Society. Now retired from teaching, he remains very active in the welding community, especially in the field of education.

Lawrence Bower is a welding instructor at Blackhawk Technical College, an AWS SENSE School, in Janesville, Wisconsin. Mr. Bower is an AWS-certified Welding Inspector and Welding Educator. In helping to create *Welding: Skills, Processes and Practices for Entry-Level Welders*, he has brought to bear an excellent mix of training experience and manufacturing know-how from his work in industry, including fourteen years at United Airlines, and six years in the US Navy as an aerospace welder.

CHAPTER 1

Gas Metal Arc Welding Equipment, Setup, and Operation

OBJECTIVES

After completing this chapter, the student should be able to

■ demonstrate the proper use of personal protective equipment (PPE) for gas metal arc welding (GMAW)

■ list the various terms used to describe gas metal arc welding

■ describe methods of metal transfer including the axial spray metal transfer process, globular transfer, pulsed-arc metal transfer (GMAW-P), and short-circuiting transfer (GMAW-S)

■ list four common shielding gases or gas mixtures used for short-circuiting, spray, and pulsed-spray transfer on plain carbon steel

■ locate the GMA welding filler metal on a welding procedure specification (WPS)

■ define *deposition efficiency,* and tell how a welder can control the deposition rate

■ define *voltage, electrical potential, amperage,* and *electrical current* as related to GMA welding

■ tell how wire-feed speed is determined and demonstrate its relationship to welding current

■ list five ways the GMAW molten weld pool can be controlled by varying the shielding gas, power settings, travel speed, electrode extension, and gun angle

■ describe and demonstrate the backhand and forehand welding techniques and their relationship to weld bead profile and penetration in the short-circuiting transfer mode

■ list and describe the basic GMAW equipment

■ use a chart to select the correct eye and face protective devices for working and welding in a shop

■ describe what type of general work clothing should be worn in a welding shop

■ describe special protective clothing worn by welders to protect hands, arms, body, waist, legs, and feet

KEY TERMS

axial spray metal transfer

electrode extension (stickout)

globular transfer

pinch effect

pulsed-arc metal transfer

short-circuiting transfer

slope

synergic systems

transition current

welding helmet

1

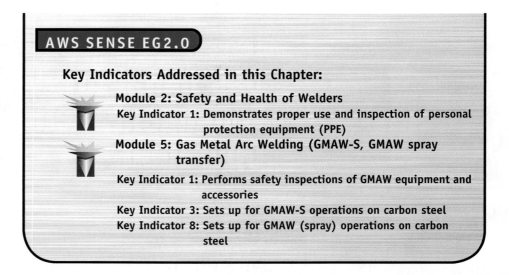

INTRODUCTION

In the 1920s, a metal arc welding process using an unshielded wire was being used to assemble the rear axle housings for automobiles. The introduction of the shielded metal arc welding electrode rapidly replaced the bare wire. The shielded metal arc welding electrode made a much higher-quality weld. In 1948, the first inert gas metal arc welding (GMAW) process, as it is known today, was developed and became commercially available, Figure 1.1. In the beginning, the GMAW process was used to weld aluminum using argon (Ar) gas for shielding. As a result, the process was known as MIG, which stands for metal inert gas welding. The later introduction of CO_2 and O_2 to the shielding gas has resulted in the American Welding Society's preferred term *gas metal*

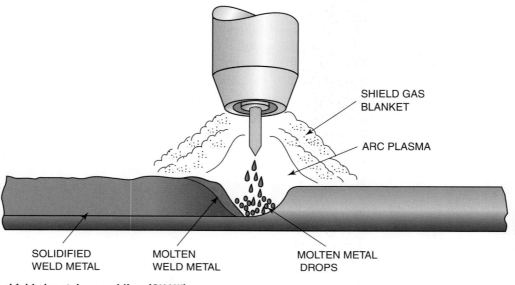

SHIELD GAS BLANKET

ARC PLASMA

SOLIDIFIED WELD METAL

MOLTEN WELD METAL

MOLTEN METAL DROPS

Figure 1.1 Gas-shielded metal arc welding (GMAW)

Table 1.1 Methods of Performing Welding Processes

Function	Manual (MA) (Example: SMAW)	Semiautomatic (SA) (Example: GMAW)	Machine (ME) (Example: GMAW)	Automatic (AU) (Example: GMAW)
Maintain the arc	Welder	Machine	Machine	Machine
Feed the filler metal	Welder	Machine	Machine	Machine
Provide the joint travel	Welder	Welder	Machine	Machine
Provide the joint guidance	Welder	Welder	Welder	Machine

arc welding (GMAW). Although the American Welding Society uses the term *gas metal arc welding* to describe this process, it is known in the field by several other terms, such as

- MIG, which is short for metal inert gas welding
- MAG, which is short for metal active gas welding
- *wire welding,* which describes the electrode used

The GMAW process may be performed as semiautomatic (SA), machine (ME), or automatic (AU) welding, Table 1.1. The GMA welding process is commonly performed as a semiautomatic process and is often mistakenly referred to as "semiautomatic welding." Equipment is available to perform most of the wire-feed processes semiautomatically, and the GMAW process can be fully automated. Robotic arc welding often uses GMAW because of the adaptability of the process in any position, Figure 1.2.

The rising use of all the various types of consumable wire welding processes has resulted in the increased sales of wire. At one time, wire made up less than 1% of the total market of filler metal. The total tonnage of filler metals used has grown and so has the percentage of wire. Today, wire exceeds 50% of the total tonnage of filler metals produced and used.

Much of the increase in the use of the wire welding processes is due to the increases in the quality of the welds produced. This improvement is due to an increased reliability of the wire-feed systems, improvements in the filler metal, smaller wire sizes, faster welding speed, higher weld deposition rates, less expensive shielding gases, and improved welding techniques. Table 1.2 shows the typical weld deposition rates using the GMA welding process. The increased usage has led to a reduction in the cost of equipment. GMA welding equipment is now found even in small shops.

In this chapter, the semiautomatic GMA welding process will be covered. The skill required to set up and operate this process is basic to the understanding and operation of other wire-feed processes. The reaction of the weld to changes in voltage, amperage, feed speed, stickout, and gas is similar to that of most wire-feed processes.

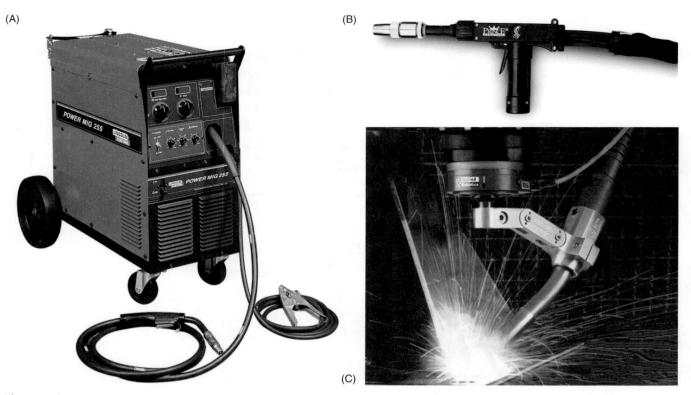

Figure 1.2 GMAW equipment
(A) Semiautomatic GMA welding setup. (B) Machine GMA welding gun with a friction drive, which provides both uniform nozzle-to-work distance and travel speed. (C) Automatic GMA welding.
Source: Courtesy of Lincoln Electric Company

Table 1.2 GMA Weld Deposition Rates

Electrode Diameter	Pounds per Hour		
Amperage	0.35	0.45	0.63
50	2.0	–	–
100	4.8	4.2	–
150	7.5	6.7	5.1
200	–	8.7	7.8
250	–	12.7	11.1
300	–	–	14.4

METAL TRANSFER

When first introduced, the GMA welding process was used with argon as a shielding gas to weld aluminum. Even though argon (Ar) was then expensive, the process was accepted immediately because it was much more productive than the gas tungsten arc (GTA) process and because it produced higher-quality welds than the shielded metal arc (SMA) process. This new arc welding process required very little postweld cleanup because it was slag and spatter free.

Axial Spray Metal Transfer

The freedom from spatter associated with the argon-shielded GMAW process results from a unique mode of metal transfer called **axial spray metal transfer**, Figure 1.3. This process is identified by the pointing of the wire tip from which droplets smaller than the diameter of the electrode wire are projected axially across the arc gap to the molten weld pool. There are hundreds of drops per second crossing from the wire to the base metal. These drops are propelled by arc forces at high velocity in the direction the wire is pointing. In the case of plain carbon steel and stainless steels, the molten weld pool may be too large and too hot to be controlled in vertical or overhead positions. Because the drops are very small and directed at the molten weld pool, the process is spatter free.

The spray transfer mode for carbon steels requires three conditions: argon-rich shielding gas mixtures, direct current electrode positive (DCEP) polarity (also called direct current reverse polarity, DCRP), and a current level above a critical amount called the **transition current**. The shielding gas is usually a mixture of 95% to 98% argon and 5% to 2% oxygen, or 80% to 90% argon and 20% to 10% carbon dioxide. The added percentage of active gases allows greater weld penetration. Figure 1.4 illustrates how the rate of drops transferred changes in relationship to the welding current. At low currents, the drops are large and are transferred at rates below 10 per second. These drops move slowly, falling from the electrode tip as gravity pulls them down. They tend to bridge the gap between the electrode

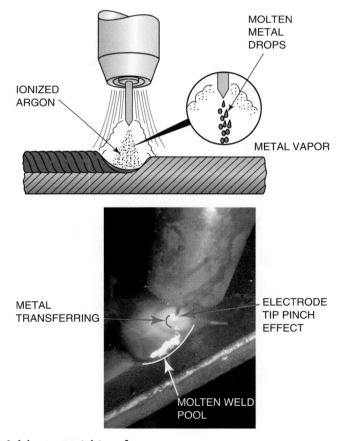

Figure 1.3 Axial spray metal transfer
Note the pinch effect of filler wire and the symmetrical metal transfer column.
Source: Courtesy of Larry Jeffus

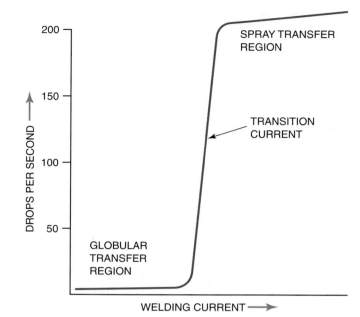

Figure 1.4 Desirable spray transfer shown schematically

tip end and the molten weld pool. This produces a momentary short circuit that throws off spatter. However, the mode of transfer changes abruptly above the critical current, producing the desirable spray.

The transition current depends on the alloy being welded. It is also proportional to the wire diameter, meaning that higher currents are needed with larger diameter wires. The need for high current density imposes some restrictions on the process. The high current hinders welding of sheet metal because the high heat cuts through sheet metal. High current also limits its use to the flat and horizontal welding positions. Weld control in the vertical or overhead position is very difficult to impossible to achieve. Table 1.3 lists the welding parameters for a variety of gases, wire sizes, and metal thicknesses for GMA welding of mild steel.

Table 1.3 GMA Welding Parameters for Mild Steel

Mild Steel Base-material Thickness, in.	Wire-feed Speed, in./min		Voltage, V				Current A
	0.035-in.	0.045-in.	CO_2	75 Ar-25 CO_2	Ar	98 Ar-2O_2	
0.036	105–115	–	18	16	–	–	50–60
0.048	140–160	70	19	17	–	–	70–80
0.060	180–220	90–110	20	17.7	–	–	90–110
0.075	240–260	120–130	20.7	18	20	–	120–130
1/8	280–300	140–150	21.5	18.5	20.5	–	140–150
3/16	320–340	160–175	22	19	21.5	23.5	160–170
1/4	360–380	185–195	22.7	19.5	22.5	24.5	180–190
5/16	400–420	210–220	23.5	20.5	23.5	25	200–210
3/8	420–520	220–270	25	22	25	26.5	220–250
1/2 and up	–	375	28	26	29	31	300

Globular Transfer

The **globular transfer process** is rarely used by itself because it transfers the molten metal across the arc in much larger droplets. It is used in combination with pulsed-spray transfer.

Globular transfer can be used on thin materials and at a very low current range. It can be used with higher current but is not as effective as other welding modes of metal transfer.

Pulsed-Spray Transfer: GMAW-P

Because the change from spray arc to globular transfer occurs within a very narrow current range and the globular transfer occurs at the rate of only a few drops per second, a controlled spray transfer at significantly lower average currents is achievable. Gas metal arc welding with pulsed-spray transfer (GMAW-P), or **pulsed-arc metal transfer** involves pulsing the current from levels below the transition current to those above it. The time interval below the transition current is short enough to prevent a drop from developing. About 0.1 second is needed to form a globule, so no globule can form at the electrode tip if the time interval at the low base current is about 0.01 second. Actually, the energy produced during this time is very low—just enough to keep the arc alive.

The arc's real work occurs during those intervals when the current pulses to levels above the transition current. The time of that pulse is controlled to allow a single drop of metal to transfer. This is typical of the drops normally associated with spray transfer. In fact, with many power supplies, a few small drops could transfer during the pulse interval. As with conventional spray arc, the drops are propelled across the arc gap, allowing metal transfer in all positions.

The average current can be reduced sufficiently to reduce penetration enough to weld sheet metal or reduce deposition rates enough to control the molten weld pool in all positions. This level controlling the weld heat input and rate of weld metal deposit is achieved by changing the following variables, graphed in Figure 1.5:

- Frequency—The number of times the current is raised and lowered to form a single pulse; frequency is measured in pulses per second.
- Amplitude—The amperage or current level of the power at the peak or maximum, expressed in amperage.
- Width of the pulses—The amount of time the peak amperage is allowed to stay on.

Figure 1.6 shows a typical pulsed-arc welding system. Although developed in the mid-1960s, this technology did not receive much attention until solid state electronics were developed to handle the high power required of welding power supplies. Solid state electronics provided a better, simpler, and more economical way to control the pulsing process. The newest generation of pulsed-arc systems interlocks the power supply and wire feeder so that the proper settings of the wire-feed and power supply are obtained for any given job by adjusting a single knob. Such systems have been termed **synergic systems**. The greatest benefit to synergic GMAW-P is that the power supply reacts to changes in the

Caution

The heat produced during pulsed-spray transfer welding using large-diameter wire or high current may be intense enough to cause the filter lens in a welding helmet to shatter. Be sure the helmet is equipped with a clear plastic lens on the inside of the filter lens. Avoid getting your face too close to the intense heat.

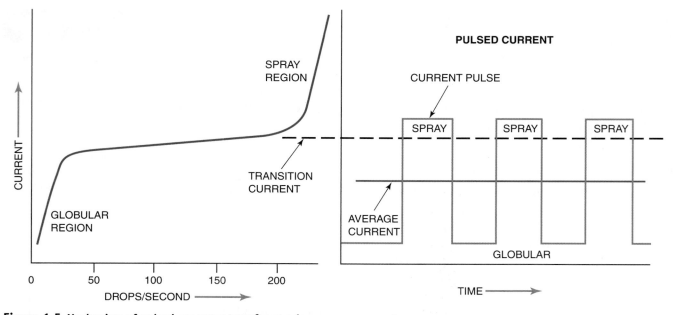

Figure 1.5 **Mechanism of pulsed-arc spray transfer at a low average current**

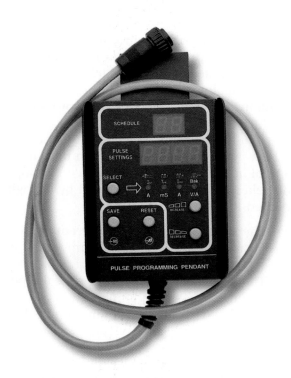

Figure 1.6 **GMA pulsed-arc welding system controller**
Source: Courtesy of Thermal Arc, a Thermadyne Company

work conditions such as inconsistent fitups or operator changes in stick-out. For example, with a traditional GMA weld, amperage is reduced as the operator increases the stickout; but with synergic GMAW-P, the power supply senses the change in stickout and adjusts the current to maintain a more consistent weld.

In some respects, synergic GMAW-P systems are more complex to use because the correct relationships between the wire-feed speeds and power supply settings must be programmed into the equipment, and each wire composition, wire size, and shielding gas requires a special program. The manufacturer generally programs the most common combinations, allowing space in the computer for additional user input. In addition, welding power supplies that produce synergic pulsation must be capable of both constant-current (CC) and constant-voltage (CV) power output.

Shielding Gases for Spray or Pulsed-spray Transfer

Axial spray transfer is impossible without shielding gases rich in argon. Pure argon is used with all metals except steels. As much as 80% helium can be added to the argon to increase the heat in the arc without affecting the desirable qualities of the spray mode. With more helium, the transfer becomes progressively more globular, forcing the use of a different welding mode, to be described later. Since these gases are inert, they do not react chemically with any metals. This factor makes the GMAW process the only productive manual or semiautomatic method for welding metals sensitive to oxygen (essentially all metals except iron or nickel). The cathodic cleaning action which helps to remove the thin layer oxides that form on metals is associated with argon at DCEP (DCRP) and is also very important for fabricating metals such as aluminum, which quickly develop these undesirable surface oxides when exposed to air.

This same cleaning action causes problems with steels. Iron oxide in and on the steel surface is a good emitter of electrons that attracts the arc. But these oxides are not uniformly distributed, resulting in very irregular arc movement and in turn irregular weld deposits. This problem was solved by adding small amounts of an active gas such as oxygen or carbon dioxide to the argon. The reaction produces a uniform film of iron oxide on the weld pool and provides a stable site for the arc. This discovery enabled uniform welds in ferrous alloys and expanded the use of GMAW to welding those materials.

The amount of oxygen needed to stabilize arcs in steel varies with the alloy. Generally, 2% is sufficient for carbon and low-alloy steels. In the case of stainless steels, about 0.5% should prevent a refractory scale of chromium oxide, which can be a starting point for corrosion in stainless steels. Carbon dioxide can substitute for oxygen. Between 8% and 10% is optimum for low-alloy steels. In many applications, carbon dioxide is the preferred addition because the weld bead has a better contour and the arc appears to be more stable. Gas mixes of 98% argon–2% oxygen as well as 98% argon–2% carbon dioxide are commonly used for spray transfer GMA welding of stainless steels.

Buried-arc Transfer

Carbon dioxide was one of the first gases studied during the development of the GMAW process. It was abandoned temporarily because of excessive spatter and porosity in the weld. After argon was accepted for shielding, further work with carbon dioxide demonstrated that the spatter was associated with globular metal transfer. The large drops are partially supported by arc forces, Figure 1.7. As they become heavy enough to overcome those forces and drop into the pool, they bridge the gap between the wire and the weld pool, producing explosive short circuits and spatter.

Additional work showed that the arc in carbon dioxide was very forceful. Because of this, the wire tip could be driven below the surface of the molten weld pool. With the shorter arcs, the drop size is reduced, and any spatter produced as the result of short circuits is trapped in the cavity produced by the arc—hence the name **buried-arc transfer**, Figure 1.8. The resultant welds tend to be more highly crowned than those produced with open arcs, but they are relatively free of spatter and offer a decided advantage of welding speed. These characteristics make the buried-arc process useful for high-speed mechanized welding of thin sections, such as that found in compressor domes for hermetic air-conditioning and refrigeration equipment or for automotive components.

Because carbon dioxide is an oxidizing gas, its applications to welding carbon steels are restricted. It cannot be used to fabricate most nonferrous materials. Neither should it be used to weld stainless steels, because carbon corrodes the weld metal.

Carbon dioxide and helium are similar in that metal transfer in both gases is globular. Helium, too, can be used with the buried-arc technique. It has the advantage of being inert, potentially making it useful for the same types of applications as carbon dioxide but in nonferrous alloys.

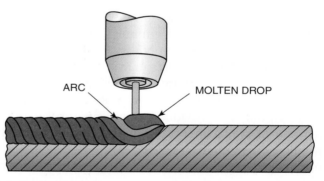

Figure 1.7 Globular metal transfer
Large drop is supported by arc forces.

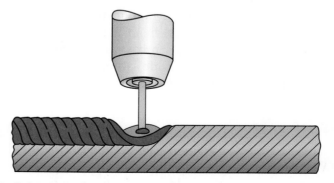

Figure 1.8 Buried-arc transfer
Wire tip is within the weld crater, so spatter is trapped.

Short-circuiting Transfer: GMAW-S

Low currents allow the liquid metal at the electrode tip to be transferred by direct contact with the molten weld pool. This process requires close interaction between the wire feeder and the power supply. This technique is called **short-circuiting transfer**, or gas metal arc welding with short-circuiting transfer (GMAW-S).

The short-circuiting mode of transfer is the most common process used with GMA welding

- on thin or properly prepared thick sections of material
- on a combination of thick to thin materials
- with a wide range of electrode diameters
- with a wide range of shielding gases
- in all positions

The 0.023, 0.030, 0.035, and 0.045 wire electrodes are the most common diameters for the short-circuiting mode. The most popular shielding gases used on carbon steel are 100% carbon dioxide (CO_2) or a combination of 75% argon (Ar) and 25% CO_2. The amperage range may be as low as 35 for materials of 24 gauge or as high as 225 for materials up to 1/8 inch in thickness on square groove weld joints. Thicker base metals can be welded if the edges are beveled to accept complete joint weld penetration.

The transfer mechanisms in this process are quite simple and straightforward, as shown schematically in Figure 1.9. To start, the wire is in direct contact with the molten weld pool, Figure 1.9A. Once the electrode touches the molten weld pool, the arc and its resistance are removed. Without the arc resistance, the welding amperage quickly rises as it begins to flow freely through the tip of the wire into the molten weld pool. The resistance to current flow is highest at the point where the electrode touches the molten weld pool. The resistance is high because both the electrode tip and weld pool are very hot. The higher the temperature,

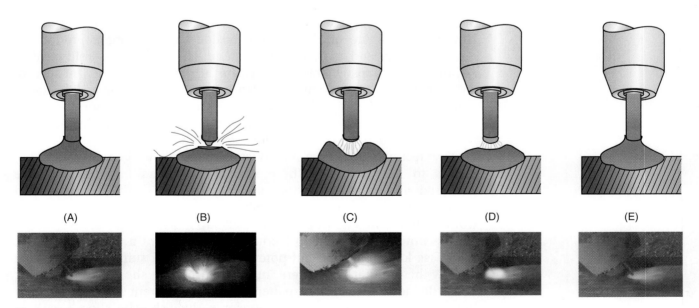

(A)   (B)   (C)   (D)   (E)

Figure 1.9 Schematic of short-circuiting transfer
Source: Courtesy of Larry Jeffus

the higher the resistance to current flow. A combination of high current flow and high resistance causes a rapid rise in the temperature of the electrode tip.

As the current flow increases, the interface between the wire and molten weld pool is heated until it explodes into a vapor (Figure 1.9B), establishing an arc. This small explosion produces sufficient force to depress the molten weld pool. A gap between the electrode tip and the molten weld pool (Figure 1.9C) immediately opens. With the resistance of the arc reestablished, the voltage increases as the current decreases.

The low current flow is insufficient to continue melting the electrode tip off as fast as it is being fed into the arc. As a result, the arc length rapidly decreases (Figure 1.9D) until the electrode tip contacts the molten weld pool (Figure 1.9E). The liquid formed at the wire tip during the arc-on interval is transferred by surface tension to the molten weld pool, and the cycle begins again with another short circuit.

If the system is properly tuned, the rate of short circuiting can be repeated from approximately 20 to 200 times per second, causing a characteristic buzzing sound. The spatter is low and the process easy to use. The low heat produced by GMAW-S makes the system easy to use in all positions on sheet metal, low-carbon steel, low-alloy steel, and stainless steel ranging in thickness from 25 gauge (0.02 in.; 0.5 mm) to 12 gauge (0.1 in.; 2.6 mm). The short-circuiting process does not produce enough heat to make quality welds in sections much thicker than 1/4 in. (6 mm) unless it is used for the root pass on a grooved weld or to fill gaps in joints. Although this technique is highly effective, lack-of-fusion defects can occur unless the process is perfectly tuned and the welder is highly skilled, especially on thicker metal. For this reason the American Welding Society D1.1 code for structural steel does not list short-circuiting GMAW as a prequalified process. In this code all short-circuiting procedures must be qualified by extensive testing.

Carbon dioxide works well with this short-circuiting process because it produces the forceful, high-energy arc needed during the arc-on interval to displace the weld pool. Helium can be used as well. Pure argon is not as effective because its arc tends to be sluggish and not very fluid. However, a mixture of 25% carbon dioxide and 75% argon produces a less harsh arc and a flatter, more fluid, and desirable weld profile. Although more costly, this gas mixture is preferred. A gas mixture of 98% argon and 2% oxygen may also be used on thinner carbon steels and sheet metal. This mixture produces lower-energy short-circuiting transfer and can be an advantage on thin-gauge metals, producing minimal burn-through at voltages as low as 13 volts.

New technology in wire manufacturing has allowed smaller wire diameters to be produced. These smaller diameters have become the preferred size even though they are more expensive due to the cost of drawing wires down to these desirable sizes. The short-circuiting process works better with a short electrode stickout.

The power supply is most critical. It must have a constant voltage, otherwise known as constant-potential output, and sufficient inductance to slow the time rate of current increase during the short-circuit interval. Too little inductance causes spatter due to high-current surges. Too much inductance causes the system to become sluggish. The short-circuiting rate decreases enough to make the process difficult to use. Also, the power supply must sustain an arc long enough to premelt the electrode tip in anticipation of the transfer at recontact with the weld pool.

FILLER METAL SPECIFICATIONS

GMA welding filler metals are available for a variety of base metals, Table 1.4. The most frequently used filler metals are AWS specification A5.18 for carbon steel and AWS specification A5.9 for stainless steel. Wire electrodes are produced in diameters of 0.023, 0.030, 0.035, 0.045, and 0.062 inch. Other, larger diameters are available for production work and can include wire diameter sizes such as 1/16, 5/64, and 7/64 inch. Table 1.5 lists the most common sizes and the amperage ranges for these electrodes. The amperage will vary depending on the method of metal transfer, type of shielding gas, and base metal thickness. Some steel wire electrodes have a thin copper coating. This coating provides some protection to the electrode from rusting and improves the electrical contact between the wire electrode and the contact tube. It also acts as a lubricant to help the wire move more smoothly through the liner and contact tube. These electrodes may look like copper wire because of the very thin copper cladding. The amount of copper is so small that it

Table 1.4 AWS Filler Metal Specifications for Different Base Metals

Base Metal Type	AWS Filler Metal Specification
Aluminum and aluminum alloys	A5.10
Copper and copper alloys	A5.6
Magnesium alloys	A5.19
Nickel and nickel alloys	A5.14
Stainless steel (austenitic)	A5.9
Steel (carbon)	A5.18
Titanium and titanium alloys	A5.16

Table 1.5 Filler Metal Diameters and Amperage Ranges

| Base Metal | Electrode Diameter | | Amperage |
	Inch	Millimeter	Range
Carbon Steel	0.023	0.6	35–190
	0.030	0.8	40–220
	0.035	0.9	60–280
	0.045	1.2	125–380
	1/16	1.6	275–450
Stainless Steel	0.023	0.6	40–150
	0.030	0.8	60–160
	0.035	0.9	70–120
	0.045	1.2	140–310
	1/16	1.6	280–450

either burns off or is diluted into the weld pool with no significant effect on the weld deposit.

Filler Metal Selection

Solid Wire

The AWS specification for carbon steel filler metals for gas-shielded welding wire is A5.18. Filler metal classified within this specification can be used for GMAW, gas tungsten arc welding (GTAW), and plasma arc welding (PAW) processes. Because in GTAW and PAW the wire does not carry the welding current, the letters *ER* are used as a prefix. The *ER* is followed by two numbers to indicate the minimum tensile strength of a good weld. The actual strength is obtained by adding three zeroes to the right of the number given. For example, ER70S-x is 70,000 psi.

The *S* located to the right of the tensile strength indicates that this is a solid wire. The last number—2, 3, 4, 5, 6, or 7—or the letter *G* is used to indicate the filler metal composition and the weld's mechanical properties, Figure 1.10.

ER70S-2

This is a deoxidized mild steel filler wire. The deoxidizers allow this wire to be used on metal that has light coverings of rust or oxides. There may be a slight reduction in the weld's physical properties if the weld is made on rust or oxides, but this reduction is only slight, and the weld will usually still pass the classification test standards. This is a general-purpose filler that can be used on killed, semikilled, and rimmed steels. Argon-oxygen, argon-CO_2, and CO_2 can be used as shielding gases. Welds can be made in all positions.

ER70S-3

This is a popular filler wire. It can be used in single or multiple-pass welds in all positions. ER70S-3 does not have the deoxidizers required to weld over rust, over oxides, or on rimmed steels. It produces high-quality welds on killed and semikilled steels. Argon-oxygen, argon-CO_2, and CO_2 can be used as shielding gases. This is the low-carbon steel filler most commonly used to weld galvanized steel.

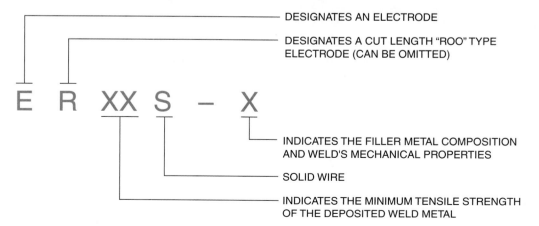

Figure 1.10 AWS numbering system for carbon steel filler metal for GMAW

ER70S-6

This is a good general-purpose filler wire. It has the highest levels of manganese and silicon. The wire can be used to make smooth welds on sheet metal or thicker sections. Welds over rust, oxides, and other surface impurities will lower the mechanical properties, but not normally below the specifications of this classification. Argon-oxygen, argon-CO_2, and CO_2 can be used as shielding gases. Welds can be made in all positions. This filler wire is not suggested for galvanized steel as its higher silicon content may induce cracking.

Stainless Steel Electrodes

The AWS specification for stainless steel–covered arc electrodes is A5.4 and for stainless steel bare, cored, and stranded electrodes and welding rods is A5.9. Filler metal classified within A5.4 uses the letter E as its prefix, and the filler metal within A5.9 uses the letters ER as its prefix.

Following the prefix, the three-digit stainless steel number from the American Iron and Steel Institute (AISI) is used. This number indicates the type of stainless steel in the filler metal, Table 1.6.

ALUMINUM AND ALUMINUM ALLOYS

The International Alloy Designation System is the most widely accepted naming scheme for aluminum alloys. Each alloy has a four-digit number, in which the first digit indicates the major alloying elements. Heat-treatable alloys can be strengthened by furnace operations after they are produced, while non-heat-treatable alloys are work hardened by plastic deformation such as bending and rolling operations. Heat treatment of aluminum alloys is sometimes referred to as precipitation hardening, and work hardening is sometimes referred to as strain hardening.

- 1000 series are essentially pure aluminum with a minimum 99% aluminum content by weight, and can be work hardened.
- 2000 series are alloyed with copper, and are heat treatable.
- 3000 series are alloyed with manganese, and can be work hardened.
- 4000 series are alloyed with silicon and can be work hardened.
- 5000 series are alloyed with magnesium, derive most of their strength from heat treatment, but can also be work hardened.
- 6000 series are alloyed with magnesium and silicon, are easy to machine, and are heat treatable, but not to the high strengths that 2000, 5000, and 7000 can reach.
- 7000 series are alloyed with zinc, and can be heat treated to the highest strengths of any aluminum alloy.
- 8000 series are a miscellaneous category where other elements not listed above may be used.

The AWS specifications for aluminum and aluminum alloy filler metals are A5.3 for covered arc welding electrodes and A5.10 for bare welding rods and electrodes. Filler metal classified within A5.3 uses the atomic symbol Al, and in A5.10 the prefix ER is used with the Aluminum Association number for the alloy, Table 1.7.

Table 1.6 Filler Metal Selector Guide for Joining Different Types of Stainless to the Same Type or Another Type of Stainless

AISI TYPE NUMBER	442 / 446	430F / 430FSE	430 / 431	501 / 502	416 / 416SE	403 / 405 / 410 / 420 / 414	321 / 348 / 347	317	316L	316	314	310 / 310S	309 / 309S	304L	303 / 303SE	201 202 301 302 302B 304 305 308	Mild Steel
201–202–301	310	310	310	310	309	309	308	308	308	308	308	308	308	308	308	308	312
302–3028–304	312	312	312	312	310	310											310
305–308	309	309	309	309	312	312											309
303	310	310	310	310	309	309	308	308	308	308	308	308	308	308	308-15		312
303SE	309	309	309	309	310	310											310
304L	310	310	310	310	309	309	308	308	308-L	308	308	308	308	308-L			312
	309	309	309	309	310	310											310
	312	312	312	312	312	312											309
309	310	310	310	310	309	309	308	308	316	316	309	309	309				309
309S	309	309	309	309	310	310	316	316			310	310	310				310
310	310	310	310	310	310	310	308	317	316	316	310	310	309				312
310S	309	309	309	309	309	309	309	316	309			310	310				310
314	310	310	310	310	310	310	309	309	309	309	310-15		309				310
	312	312	312	312	312	312	310	310	310	310	15		310				309
316	310	310	310	310	309	309	308	316	316	316	309	310	309		308	309	310
	309	309	309	309	310	310	310	316	316	310	310	309	310	316	316	310	310
	312	312	312	312	312	312	308			316	316	316	316				312

This page contains a large rotated cross-reference chart for welding filler metal selection. The recommended filler metals (first, second, and third choices) are listed for combinations of base metals.

Base Metal	Recommended Filler (1st / 2nd / 3rd choice) across cross-reference columns
316L	310/309/312 · 310/309/312 · 310/309/312 · 310/309/312 · 310/309/312 · 309/310/312 · 309/310/312 · 308 · 316/317/308 · 316 · 309/310 · 309/310/312 · 308/316 · 308/316 · 308/316 · 309/310/312
317	310/309/312 · 310/309/312 · 310/309/312 · 310/309/312 · 310/309/312 · 309/310/312 · 309/310/312 · 317/308 · 317/316/309 · 316/308 · 316 · 310/316 · 317 · 316 · 316/309/317 · 316/316/317 · 310/316
321	310/309/312 · 310/309/312 · 310/309/312 · 310/309/312 · 310/309/312 · 309/310/312 · 309/310/312 · 309 · 308/347 · 347 · 347 · 347/308 · 347/308 · 347/308 · 347/308 · 347/308
348	310/309/312 · 310/309/312 · 310/309/312 · 310/309/312 · 310/309/312 · 309/310/312 · 309/310/312 · 347 · 308 · 308 · 308 · 310/308 · 308 · 308 · 308 · 308
347	312 · 312 · 312 · 312 · 312 · 312 · 312 · — · 347 · –L · 347 · 312
403–405	310/309/312 · 310/309/312 · 310/309/312 · 310/309/312 · 410* · 309/309** · 309 · 309 · 309 · 309 · 309 · 309 · 309 · 309 · 309 · 309
410–420	309/309/312 · 309/309/312 · 310/310/312 · 310/310/312 · 309**/310** · 310 · 310 · 310 · 310 · 310 · 310 · 310 · 310 · 310 · 310 · 310
414	312 · 312 · 312 · 312 · 312 · 312 · 312 · 312 · 312 · 312 · 312 · 312 · 312
416	310/310/312 · 310/310/312 · 410–15* · 410–15* · 309 · 309 · 309 · 309 · 309 · 309 · 309 · 309 · 309 · 309 · 309
416SE	309/309 · 309 · 309* / 310** · 310 · 310 · 310 · 310 · 310 · 310 · 310 · 310 · 310
501	310 · 310 · 502* / 310** · 310 · 310 · 310 · 310 · 310 · 310 · 310 · 310 · 310 · 310
502	310** · 309 · 309 · 309 · 309 · 309 · 309 · 309 · 309 · 309 · 309
430	310/309/312 · 430–15* / 310**/309** · 310 · 310 · 310 · 310 · 310 · 310 · 310 · 310 · 310 · 310
431	309 · 309 · 309 · 309 · 309 · 309 · 309 · 309 · 309 · 309 · 309
430F	310/309/312 · 410– · 310 · 310 · 309 · 309 · 309 · 309 · 309 · 309 · 309
430FSE	309 · 15* · 309 · 309 · 312 · 312 · 312 · 312 · 312 · 312 · 312
442	309/310/312 · 309/310/312 · 309/310/312 · 310/309/312 · 310/309 · 310/309 · 310/309 · 310/309 · 310/309
443	310/312 · 310/312 · 312 · 312 · 312 · 312 · 312 · 312 · 312 · 312 · 312

Note: Bold numbers indicate first choice; light numbers indicate second and third choices. This choice can vary with specific applications and individual job requirements.

*Preheat.

**No preheat necessary.

Source: Courtesy of Thermacote Welco

Table 1.7 Recommended Filler Metals for Joining Different Types of Aluminum to the Same Type or a Different Type of Aluminum

Base Metal	319 355	43 356	214	6061 6063 6151	5456	5454	5154 5254	5086	5083	5052 5652	5005 5050	3004	1100 3003	1060
1060	4145, 4043, 4047	4043, 4047, 4145	4043	4043	5356	4043	4043	5356	5356	4043	1100	4043	1100, 4043	1260, 4043, 1100
1100	4145, 4043, 4047	4043, 4047, 4145	4043, 5183	4043, 4047	5356, 4043	4043, 5183	4043, 5183	5356, 4043	5356, 4043	4043, 4047	4043	4043	1100, 4043	
3003	4043, 4047	4043, 4047, 4145	4043, 5183	4043, 4047	5356, 4043	4043, 5183	4043, 5183	5356, 4043	5356, 4043	4043, 4047	4043, 5183, 5356	4043, 5183, 5356	4043	
3004	4043, 4047	4043, 4047	5654, 5183, 5356	4043, 5183, 5356	5356, 5183, 5556	4043, 5183, 5356	4043, 5183, 5356	5356, 5183, 5556	5356, 5183, 5556	4043, 5183, 4047	4043, 5183, 5356	4043, 5183, 5356		
5005	4043, 4047	4043, 4047	5654, 5183	4043, 5183	5356, 5183	5654, 5183	5654, 5183	5356, 5183	5356, 5183	4043, 5183	4043, 5183			
5050	4043, 4047	5183	5183	5183, 5356	5556	5183, 5356	5183, 5356	5183, 5556	5183, 5556	5183, 4047	4043, 5183			
5052	4043, 4047	4043, 5183	5654, 5183	4043, 5183	5356, 5183	5654, 5183	5654, 5183	5356, 5183	5356, 5183	4043, 5654				
5652		5183	5183	5183	5183	5183	5183	5183	5183	5183				
5083	NR	5356, 5183	5356, 5183	5356, 5183	5356	5356	5356	5356	5356					
5086	NR	5356, 5183	5356, 5183	5356, 5183	5356	5356	5356	5356						
5154	NR	5356, 5183	5654, 5183	5356, 5183	5556	5654, 5554	5654							
5254		4047	5356	5183	5183	5183	5356							
5454	4043, 4047	4043, 5183, 4047	5654, 5183, 5356	5356, 5183, 4043	5356, 5183, 5554	5554								

Table 1.7 (Continued)

Base Metal	319 355	43 356	214	6061 6063 6151	5456	5454	5154 5254	5086	5083	5052 5652	5005 5050	3004	1100 3003	1060
		5356	5356	5356	5556									
5456	NR	4043	5183	5183	5183									
		5183	5556	5556	5356									
6061	4145	4043	5356	4043										
6063	4043	5183	5183	5183										
6151	4047	4047	4043	4047										
214		4043	5654											
	NR	5183	5183											
		4047	5356											
43	4043	4145												
356	4047	4043												
		4047												
319	4145													
355	4043													
	4047													

Note: First filler alloy listed in each group is the all-purpose choice. NR means that these combinations of base metals are not recommended for welding.
Courtesy of Thermacote Welco

Aluminum Bare Welding Rods and Electrodes

ER1100

1100 aluminum has the lowest percentage of alloy agents of all the aluminum alloys, and it melts at 1215°F. The filler wire is also relatively pure. ER1100 produces welds that have good corrosion resistance and high ductility, with tensile strengths ranging from 11,000 to 17,000 psi. The weld deposit has a high resistance to cracking during welding. This wire can be used with oxyfuel gas welding (OFW), GTAW, and GMAW. Preheating to 300°F to 350°F is required for GTA welding on plate or pipe 3/8 in. and thicker to ensure good fusion. Flux is required for OFW. 1100 aluminum is commonly used for items such as food containers, food-processing equipment, storage tanks, and heat exchangers. ER1100 can be used to weld 1100 and 3003 grade aluminum.

ER4043

ER4043 is a general-purpose welding filler metal. It has 4.5% to 6.0% silicon added, which lowers its melting temperature to 1155°F. The lower melting temperature helps promote a free-flowing molten weld pool. The welds have high ductility and a high resistance to cracking during welding. This wire can be used with OFW, GTAW, and GMAW. Preheating to 300°F to 350°F is required for GTA welding on plate or pipe 3/8 in. and thicker to ensure good fusion. Flux is required for OFW. ER4043 can be used to weld on 2014, 3003, 3004, 4043, 5052, 6061, 6062, and 6063 and cast alloys 43, 355, 356, and 214.

ER5356

ER5356 has 4.5% to 5.5% magnesium added to improve the tensile strength. The weld has high ductility but only an average resistance to cracking during welding. This wire can be used for GTAW and GMAW. Preheating to 300°F to 350°F is required for GTA welding on plate or pipe 3/8 in. and thicker to ensure good fusion. ER5356 can be used to weld on 5050, 5052, 5056, 5083, 5086, 5154, 5356, 5454, and 5456.

ER5556

ER5556 has 4.7% to 5.5% magnesium and 0.5% to 1.0% manganese added to produce a weld with high strength. The weld has high ductility and only average resistance to cracking during welding. This wire can be used for GTAW and GMAW. Preheating to 300°F to 350°F is required for GTA welding on plate or pipe 3/8 in. and thicker to ensure good fusion. ER5556 can be used to weld on 5052, 5083, 5356, 5454, and 5456.

WIRE MELTING AND DEPOSITION RATES

The wire melting rates, deposition rates, and wire-feed speeds of the consumable wire welding processes are affected by the same variables. Before discussing them, however, these terms need to be defined. The wire melting rate, measured in inches per minute (in/min) or pounds per hour (lb/hr), is the rate at which the arc consumes the wire. The deposition rate, the measure of weld metal deposited, is nearly always less than the melting rate because not all of the wire is converted to weld metal. Some is lost as slag, spatter, or fume. The amount of weld metal deposited in ratio to the wire used is called the deposition efficiency.

Deposition efficiencies depend on the process, on the gas used, and even on how the welder sets welding conditions. With efficiencies of approximately 98%, solid wires with argon-rich shield gas mixes are best. Some of the self-shielded cored wires are poorest, with efficiencies as low as 80%.

Welders can control the deposition rate by changing the current, electrode extension, and diameter of the wire. To obtain higher melting rates, they can increase the current or wire extension or decrease the wire diameter. Knowing the precise constants is unimportant. However, it is important to know that current greatly affects melting rate and that the electrode extension must be controlled if results are to be reproducible.

WELDING POWER SUPPLIES

To better understand the terms used to describe the different welding power supplies, you need to know the following electrical terms:

- Voltage, or volts (V), is a measurement of electrical pressure and is the force that causes the current (amperage) to flow, in the same way that pounds per square inch is a measurement of water pressure.
- Electrical potential means the same thing as voltage and is usually expressed by using the term *potential (P)*. The terms *voltage, volts,* and *potential* can all be interchanged when referring to electrical pressure.
- Amperage, or amps (A), is the measurement of the total number of electrons flowing, in the same way that gallons are a measurement of the amount of water flowing.
- Electrical current means the same thing as amperage and is usually expressed by using the term *current (C)*. The terms *amperage, amps,* and *current* can all be interchanged when referring to electrical flow.

GMAW power supplies are constant-voltage, constant-potential (CV, CP) machines, unlike shielded metal arc welding (SMAW) power supplies, which are constant-current (CC) machines and are sometimes called drooping arc voltage (DAV). It is impossible to make acceptable welds using the wrong type of power supply. GMAW power supplies are available as transformer rectifiers, motor generators, or inverters, Figure 1.11. Some newer machines use electronics, enabling them to supply both types of power at the flip of a switch; they may be referred to as CC/CV power supplies.

The relationships between current and voltage with different combinations of arc length or wire-feed speeds are called volt-ampere characteristics. The volt-ampere characteristics of arcs in argon with constant arc lengths or constant wire-feed speeds are shown in Figure 1.12. To maintain a constant arc length while increasing current, it is necessary to increase voltage. For example, with a 1/8-in. (3-mm) arc length, increasing current from 150 to 300 amperes requires increasing the

Figure 1.11 Transformer rectifier welding power supply
Source: Courtesy of ESAB Welding & Cutting Products

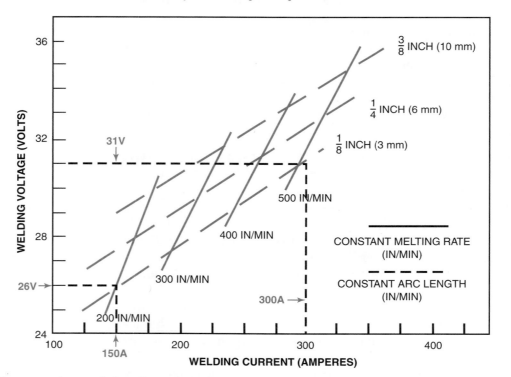

Figure 1.12 Volt-ampere characteristics of arcs with argon
The arc length and arc voltage are affected by the welding current and wire-feed speed (0.045-in. [1.43-mm] wire; 1-in. [25-mm] electrode extension).

voltage from about 26 to 31 volts. The current increase illustrated here results from increasing the wire-feed speed from 200 to 500 inches per minute.

Speed of the Wire Electrode

The wire-feed speed is generally recommended by the electrode manufacturer and is selected in inches per minute (ipm), or how fast the wire exits the contact tube. The welder uses a wire-speed control dial on the wire-feed unit to control ipm. It can be advanced or slowed to control the burn-off rate, or how fast the electrode transfers into the weld pool, to meet the welder's skill in controlling the weld pool. Note the direct relationship between current (amps) and wire-feed speed (wfs); as wire-feed speed increases, the amperage increases. If wire-feed speed is reduced, amperage will decrease, Table 1.8.

To accurately measure wire-feed ipm, snip off the wire at the contact tube. Wearing safety glasses and pointing the contact tube away from your face, squeeze the trigger for ten seconds; release and snip off the wire electrode. Measure the number of inches of wire that was fed out in the ten seconds. Now using basic shop math, multiply its total length in inches by six. The result is how many inches of wire were fed per minute.

Power Supplies for Short-circuiting Transfer

Although the GMA power source is said to have a constant voltage (CV) or constant potential (CP), it is not perfectly constant. The graph in Figure 1.13 shows that there is a slight decrease in voltage as the amperage increases within the working range. The rate of decrease is known as **slope**. It is expressed as the voltage decrease per 100-ampere increase—for example, 10 V/100 A. For short-circuiting welding, some welding power supplies are equipped to allow changes in the slope by steps or continuous adjustment.

The slope, which is called the *volt-ampere curve*, is often drawn as a straight line because it is fairly straight within the working range of the machine. Whether it is drawn as a curve or a straight line, the slope can be found by finding two points. The first point is the set voltage as read from the voltmeter when the gun switch is activated but no welding is

Table 1.8 Typical Amperages for Carbon Steel

Wire-feed Speed* in./min (m/min)	Wire Diameter Amperages			
	.030 in. (0.8 mm)	.035 in. (0.9 mm)	.045 in. (1.2 mm)	.062 in. (1.6 mm)
100 (2.5)	40	65	120	190
200 (5.0)	80	120	200	330
300 (7.6)	130	170	260	425
400 (10.2)	160	210	320	490
500 (12.7)	180	245	365	–
600 (15.2)	200	265	400	–
700 (17.8)	215	280	430	–

*To check feed speed, run out wire for one minute and then measure its length.

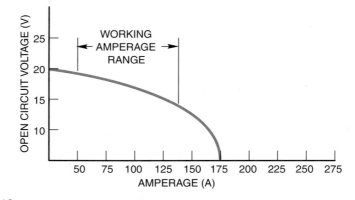

Figure 1.13 Constant potential welder slope

being done. This is referred to as the open circuit voltage. The second point is the voltage and amperage as read during a weld. The voltage control is not adjusted during the test but the amperage can vary. The slope is the voltage difference between the first and second readings. The difference can be found by subtracting the second voltage from the first voltage. Therefore, for settings over 100 amperes, it is easier to calculate the slope by adjusting the wire feed so that you are welding with 100 amperes, 200 amperes, 300 amperes, and so on. In other words, the voltage difference can be simply divided by 1 for 100 amperes, 2 for 200 amperes, and so forth.

The machine slope is affected by circuit resistance. Circuit resistance may result from a number of factors, including poor connections, long leads, or a dirty contact tube. A higher resistance means a steeper slope. In short-circuiting machines, increasing the inductance increases the slope. This increase slows the current's rate of change during short circuiting and the arcing intervals, Figure 1.14. Therefore, slope and inductance become synonymous in this discussion. As the slope increases, both the short-circuit current and the **pinch effect** are reduced. A flat slope has both an increased short-circuit current and a greater pinch effect.

The machine slope affects the short-circuiting metal transfer mode more than it does the other modes. Too much current and pinch effect from a flat slope cause a violent short and arc restart cycle, which results in increased spatter. Too little current and pinch effect from a steep slope result in the short circuit not being cleared as the wire freezes in the molten pool and piles up on the work, Figure 1.15.

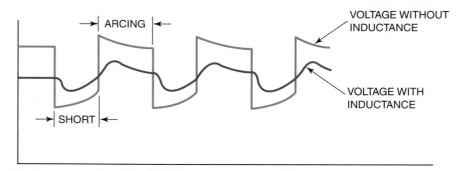

Figure 1.14 Voltage pattern with and without inductance

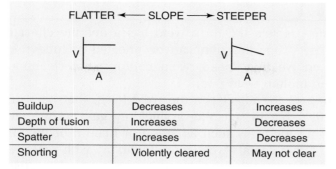

FLATTER ← SLOPE → STEEPER		
Buildup	Decreases	Increases
Depth of fusion	Increases	Decreases
Spatter	Increases	Decreases
Shorting	Violently cleared	May not clear

Figure 1.15 Effect of slope

The slope should be adjusted so that a proper spatter-free metal transfer occurs. On machines that have adjustable slopes, this is easily set. Experiment 2-2 describes a method of adjusting the circuit resistance to change the slope on machines that have a fixed slope. This is done by varying the contact tube-to-work distance. The GMA filler wire is much too small to carry the welding current and heats up due to its resistance to the current flow. The greater the tube-to-work distance, the greater the circuit resistance and the steeper the slope. By increasing or decreasing this distance, a proper slope can be obtained so that the short circuiting is smoother with less spatter.

Inductance

Many newer GMA welding power supplies are supplied with an inductance control. Inductance controls are used primarily in the GMA short-circuiting transfer mode, especially when open root joints are involved. Inductance controls the rise up to peak current during GMAW short-circuiting transfer. A low inductance control setting will provide a greater short-circuiting frequency, which may be beneficial in welding thin material where burn-through is an issue. A higher setting will reduce the frequency of short circuits, creating slightly longer arc periods, which allows greater current to go to the work. The greater current can be used to increase penetration in thicker sections or where complete penetration is required on open root joints welded from one side.

In the case of the GMA welding arc, the rate of change in the amperage relative to the arc voltage determines how the metal droplet detaches from the end of the electrode. If the rate of change is too rapid (inductance too low), the droplet detaches violently and produces excessive spatter. If the rate of change is too slow (inductance too high) the metal droplet doesn't detach cleanly and the arc is unstable. A midrange setting on the inductance control can be used for general-purpose short-circuit GMA welding and adjustments made for thick, thin, or open root conditions.

MOLTEN WELD POOL CONTROL

The GMAW molten weld pool can be controlled by varying the following factors: shielding gas, power settings, gun manipulation, travel speed, electrode extension, and gun angle.

Shielding Gas

The shielding gas selected for a weld has a definite effect on the weld produced. The properties that can be affected include the method of metal transfer, welding speed, weld contour, arc cleaning effect, and fluidity of the molten weld pool.

In addition to the effects on the weld itself, the metal to be welded must be considered in selecting a shielding gas. Some metals must be welded with an inert gas such as argon or helium or mixtures of argon and helium. Other metals weld more favorably with reactive gases such as carbon dioxide or with mixtures of inert gases and reactive gases such as argon and oxygen or argon and carbon dioxide, Table 1.9. The most commonly used shielding gases are 75% argon + 25% CO_2, argon + 1% to 5% oxygen, and carbon dioxide, Figure 1.16.

- **Argon:** The atomic symbol for argon is *Ar*, and it is an inert gas. *Inert gases* do not react with any other substance and are insoluble in molten metal. One hundred percent argon is used on nonferrous metals such as aluminum, copper, magnesium, nickel, and their alloys; but 100% argon is not normally used for making welds on ferrous metals.

  Because argon is denser than air, it effectively shields welds by pushing the lighter air away. Argon is relatively easy to ionize. Easily ionized gases can carry long arcs at lower voltages. This makes it less sensitive to changes in arc length.

  Argon gas is naturally found in all air and is collected in air separation plants. There are two methods of separating air to extract nitrogen, oxygen, and argon. In the cryogenic process, air is super-cooled to temperatures that cause it to liquefy and the gases are separated. In the noncryogenic process, molecular sieves (strainers with very small holes) separate the various gases, much like using a screen to separate sand from gravel.

- **Argon gas blends:** Oxygen, carbon dioxide, helium, and nitrogen can be blended with argon to change argon's welding characteristics. Adding reactive gases (oxidizing), such as oxygen or carbon dioxide, to argon tends to stabilize the arc, promote favorable metal transfer, and minimize spatter. As a result, the penetration pattern is improved, and undercutting is reduced or eliminated. Adding helium or nitrogen gases (nonreactive or inert) increases the arc heat for deeper penetration.

  The amount of the reactive gases, oxygen or carbon dioxide, required to produce the desired effects is quite small. As little as a half percent change in the amount of oxygen will produce a noticeable effect on the weld. Most of the time, blends containing 1% to 5% of oxygen are used. Carbon dioxide may be added to argon in the range of 10% to 30%. Blends of argon with less than 10% carbon dioxide may not have enough arc voltage to give the desired results. The most commonly used blend for short-circuiting transfer is 75% argon and 25% CO_2.

  When using oxidizing shielding gases with oxygen or carbon dioxide added, a suitable filler wire containing deoxidizers should be used to prevent porosity in the weld. The presence of oxygen in the shielding gas can also cause some loss of certain alloying

Table 1.9 (A) Metals Matched with GMAW Shielding Gases and Gas Blends; (B) GMAW Shielding Gases and Gas Blends Matched with Metals

GMAW Metals, Shielding Gases, and Gas Blends

Metals	Gases		Blends of Two Gases									Blends of Three Gases		
			Argon + Oxygen			Argon + Carbon Dioxide			Argon + Helium					
	Argon (Ar)	CO_2	Ar+ 1% O_2	Ar+ 2% O_2	Ar+ 5% O_2	Ar+ 5% CO_2	Ar+ 10% CO_2	Ar+ 25% CO_2	Ar+ 25% He	Ar+ 50% He	Ar+ 75% He	Ar+CO_2 +O_2	Ar+CO_2+ Nitrogen	Ar+CO_2+ Helium
Aluminum	X								X	X	X			X
Copper Alloys	X								X	X	X			
Stainless Steel		X	X	X							X	X	X	X
Steel		X	X	X	X	X	X	X	X			X		
Magnesium	X								X	X	X			
Nickel Alloys	X								X	X	X			

(A)

GMAW Shielding Gas, Gas Blends, Metals, and Welding Process

Gases/ Blend	Gas Reaction	Application	Remarks
Argon (Ar)	Inert	Nonferrous metals	Provides spray transfer
Helium (He)	Inert	Aluminum and magnesium	Very hot arc for welds on thick sections, usually used in gas blends to increase the arc temperature and penetration
Ar + 1% O_2	Oxidizing	Stainless steel	Oxygen provides arc stability
Ar + 2% O_2	Oxidizing	Stainless steel	Oxygen provides arc stability
Ar + 5% O_2	Oxidizing	Mild and low-alloy steel	Provides spray transfer
CO_2	Oxidizing	Mild, low-alloy steels and stainless steel	Least expensive gas, deep penetration with short-circuiting or globular transfer
Nitrogen	Almost inert	Copper and copper alloys	Has high heat input with globular transfer
Ar + 25% He	Inert	Al, Mg, copper, nickel, and their alloys	Higher heat input than Ar, for thicker metal
Ar + 50% He	Inert	Al, Mg, copper, nickel, and their alloys	Higher heat in arc use on heavier thickness with spray transfer
Ar + 75% He	Inert	Copper, nickel, and their alloys	Highest heat input
	Oxidizing	Low-alloy steel	

Table 1.9 (A) Metals Matched with GMAW Shielding Gases and Gas Blends; (B) GMAW Shielding Gases and Gas Blends Matched with Metals

Gases/Blend	Gas Reaction	Application	Remarks
Ar + 5% CO_2	Oxidizing	Low-alloy steel	Pulse spray and short-circuit transfer in out-of-position welds
Ar + 10% CO_2	Oxidizing	Low-alloy steel	Same as above with a wider, more fluid weld pool
Ar + 25% CO_2	Oxidizing	Mild, low-alloy steels and stainless steel	Smooth weld surface, reduces penetration with short-circuiting transfer

GMAW Shielding Gas, Gas Blends, Metals, and Welding Process

Gas/Blend	Gas Reaction	Application	Remarks
Ar + CO_2 + O_2	Oxidizing	Low-alloy steel and some stainless steels	All metal transfer for automatic and robotic applications
Ar + CO_2 + N	Almost inert	Stainless steel	All metal transfer, excellent for thin gauge material
He + 7.5% Ar + 2.5% CO_2	Almost inert	Stainless steel and some low-alloy steels	Excellent toughness, arc stability, wetting characteristics, and bead contour, little spatter with short-circuiting transfer

(B)

ARGON + OXYGEN   ARGON + CO$_2$ CARBON DIOXIDE

Figure 1.16 Effect of shielding gas on weld bead shape

elements, such as chromium, vanadium, aluminum, titanium, manganese, and silicon.

- **Helium:** The atomic symbol for helium is *He*, and it is an inert gas that is a product of the natural gas industry. It is removed from natural gas as the gas undergoes separation (fractionation) for purification or refinement.

  Helium is lighter than air; thus its flow rates must be about twice as high as argon's for acceptable stiffness in the gas stream to be able to push air away from the weld. Proper protection is difficult in drafts unless high flow rates are used. It requires a higher voltage to ionize, which produces a much hotter arc. There is a noticeable increase in both the heat and temperature of a helium arc. This hotter arc makes it easier to make welds on thick sections of aluminum and magnesium.

  Small quantities of helium are blended with heavier gases. These blends take advantage of the heat produced by the lightweight helium and weld coverage by the other heavier gas. Thus, each gas is contributing its primary advantage to the blended gas.

- **Carbon dioxide:** Carbon dioxide is a compound made up of one carbon atom (C) and two oxygen atoms (O$_2$), and its molecular formula is *CO$_2$*. One hundred percent carbon dioxide is widely used as a shielding gas for GMA welding of steels. In the short-circuiting transfer mode it allows higher welding speed, better penetration, good mechanical properties, and costs less than the inert gas mixes. The chief drawback in the use of carbon dioxide is the less-steady arc characteristics and a considerable increase in weld spatter. The spatter can be kept to a minimum by maintaining a very short, uniform arc length and strict attention to amperage and voltage parameters. CO$_2$ can produce sound, spatter-free welds of the highest quality, provided established procedures are followed and a filler wire having the proper deoxidizing additives is selected.

- **Nitrogen:** The atomic symbol for nitrogen is *N*. It is not an inert gas but is relatively nonreactive to the molten weld pool. It is often used in blended gases to increase the arc's heat and temperature. One hundred percent nitrogen can be used to weld copper and copper alloys and is an economical choice for gas purging of some austenitic stainless steel pipe welds.

Power Settings

As the power settings, voltage, and amperage are adjusted, the weld bead is affected. Making an acceptable weld requires a balancing of the voltage and amperage. If either or both are set too high or too low, the

 Module 5
Key Indicator 3, 8

weld penetration can decrease. A GMA welding machine has no direct amperage settings. Instead, the amperage at the arc is adjusted by changing the wire-feed speed. As a result of the welding machine's maintaining a constant voltage when the wire-feed speed increases, more amperage flows across the arc. This higher amperage is required to melt the wire so that the same arc voltage can be maintained. The higher amperage is used to melt the filler wire and does not increase the penetration. In fact, the weld penetration may decrease significantly.

Increasing and decreasing the voltage changes the arc length; however, it may not put more heat into the weld. Like changes in the amperage, these voltage changes may decrease weld penetration.

Gun Manipulation

The GMA welding process is greatly affected by the location of the electrode tip and molten weld pool. During the short-circuiting process if the arc is directed to the base metal and outside the molten weld pool, the welding process may stop. Without the resistance of the hot molten metal, high-amperage surges occur each time the electrode tip touches the base metal, resulting in a loud pop and a shower of sparks. It is something that occurs each time a new weld is started. So when making a weave pattern, which is a gun manipulation technique of moving side to side in order to produce a wider weld bead, you must keep the arc and electrode tip directed into the molten weld pool. Other than the sensitivity to arc location, most of the SMAW weave patterns that keep the electrode wire at or near the leading edge of the weld pool can be used for short-circuiting GMA welds.

Travel Speed

Because the location of the arc inside the molten weld pool is important, the welding travel speed cannot exceed the ability of the arc to melt the base metal. Too high a travel speed can result in overrunning of the weld pool and an uncontrollable arc. Fusion between the base metal and filler metal can completely stop if the travel rate is too fast. If the travel rate is too slow and the weld pool size increases excessively, it can also restrict fusion to the base plate.

Electrode Extension

Module 5
Key Indicator 3, 8

The **electrode extension (stickout)** is the distance from the contact tube to the arc measured along the wire. Adjustments in this distance cause a change in the wire resistance and the resulting weld bead, Figure 1.17.

GMA welding currents are relatively high for the wire sizes, even for the low current values used in short-circuiting arc metal transfer, Figure 1.18. As the length of wire extending from the contact tube to the work increases, the voltage, too, should increase. Since this change is impossible with a constant-voltage power supply, the system compensates by reducing the current. In other words, by increasing the electrode extension and maintaining the same wire-feed speed, the current has to change to provide the same resistance drop. This situation leads to a reduction in weld heat, penetration, and fusion, and an increase in buildup. On the

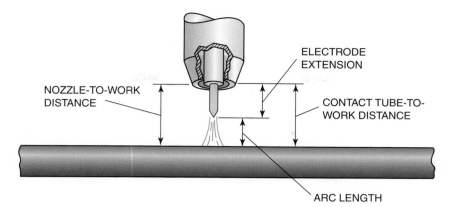

Figure 1.17 Electrode-to-work distances

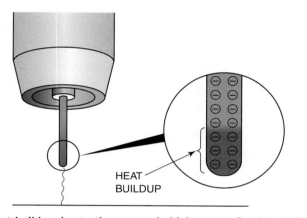

Figure 1.18 Heat buildup due to the extremely high current for the small conductor (electrode)

other hand, as the electrode extension distance is shortened, the weld heats up, penetrates more, and builds up less, Figure 1.19.

Experiment 2-3 explains the technique of using varying extension lengths to change the weld characteristics. Using this technique, a welder can make acceptable welds on metal ranging in thickness from 16 gauge to 1/4 in. (6 mm) or more without changing the machine settings. When using this technique, the nozzle-to-work distance should be kept the same so that enough shielding gas coverage is provided. Some nozzles can be extended to provide coverage. Others must be exchanged with the correct-length nozzle, Figure 1.20.

Gun Angle

Module 5
Key Indicator 3, 8

The GMA welding gun may be held so that the relative angle between the gun, work, and welding bead being made is either vertical or has a drag angle or a push angle. Changes in this angle will affect the weld bead. The effect is most noticeable during the short-circuiting arc and globular transfer modes.

Backhand welding is the welding technique that uses a drag angle, Figure 1.21. The welding technique that uses a push angle is known as forehand welding, Figure 1.22.

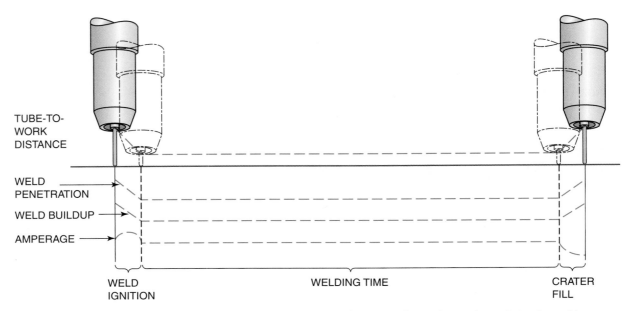

TUBE-TO-
WORK
DISTANCE

WELD
PENETRATION

WELD BUILDUP

AMPERAGE

WELD
IGNITION

WELDING TIME

CRATER
FILL

Figure 1.19 Using the changing tube-to-work distance to improve both the starting and stopping points of a weld

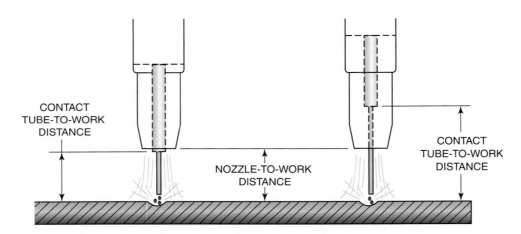

CONTACT
TUBE-TO-WORK
DISTANCE

NOZZLE-TO-WORK
DISTANCE

CONTACT
TUBE-TO-WORK
DISTANCE

Figure 1.20 Nozzle-to-work distance can differ from the contact tube-to-work distance

Figure 1.21 Backhand welding, or drag angle
Source: Courtesy of Larry Jeffus

Figure 1.22 Forehand welding, or push angle
Source: Courtesy of Larry Jeffus

Backhand Welding

A dragging angle, or backhand, welding technique directs the arc force into the molten weld pool of metal. This action, in turn, forces the molten metal back onto the trailing edge of the molten weld pool and exposes more of the unmelted base metal, Figure 1.23. The digging action pushes the penetration deeper into the base metal while building up the weld head. If the weld is sectioned, the profile of the bead is narrow and deeply penetrated, with high buildup.

Module 5
Key Indicator 3, 8

Forehand Welding

In a push angle, or forehand, welding technique, the arc force pushes the weld metal forward and out of the molten weld pool onto the cooler metal ahead of the weld, Figure 1.24. The heat and metal are spread out over a wider area. The sectional profile of the bead is wide, showing shallow penetration with little buildup.

The greater the angle, the more defined is the effect on the weld. As the angle approaches vertical, the effect is reduced. This allows the welder to change the weld bead as effectively as the changes resulting from adjusting the machine current settings.

Module 5
Key Indicator 3, 8

EQUIPMENT

The basic GMAW equipment consists of the gun, electrode (wire) feed unit, electrode (wire) supply, power source, shielding gas supply with flowmeter/regulator, control circuit, and related hoses, liners, and cables, Figure 1.25 and Figure 1.26. Larger, more complex systems may have water for cooling, solenoids for controlling gas flow, and carriages for moving the work or the gun or both, Figure 1.27. The system may be

Module 5
Key Indicator 3, 8

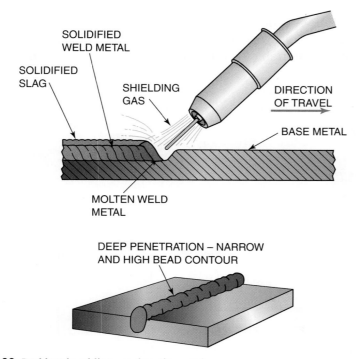

SOLIDIFIED
WELD METAL

SOLIDIFIED
SLAG

SHIELDING
GAS

DIRECTION
OF TRAVEL

BASE METAL

MOLTEN WELD
METAL

DEEP PENETRATION – NARROW
AND HIGH BEAD CONTOUR

Figure 1.23 Backhand welding, or dragging angle

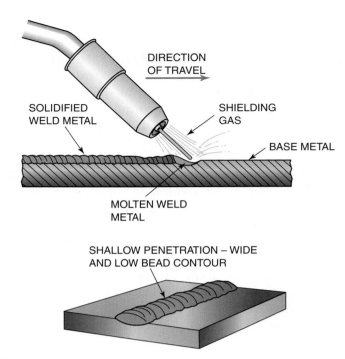

Figure 1.24 Forehand welding, or pushing angle

stationary or portable, Figure 1.28. In most cases, the system is meant to be used for only one process. Some manufacturers, however, do make power sources that can be switched over for other uses.

Power Source

The power source may be either a transformer rectifier, inverter, or generator type. The transformers are stationary and commonly require a three-phase power source. The inverter power sources are smaller, lighter, and may be designed to accept a variety of different electrical inputs, from 208 volts to 440 volts, single or three-phase. Engine generators are ideal for portable use or where sufficient power is not available.

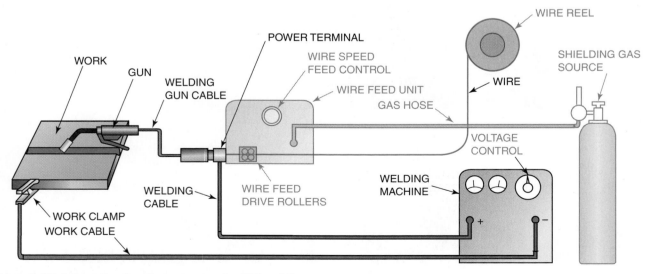

Figure 1.25 Schematic of equipment setup for GMA welding
Source: Courtesy of Hobart Brothers Company

Figure 1.26 Small 110-V GMA welder
Source: Courtesy of Thermal Arc, a Thermadyne Company

Figure 1.27 Robot welding
Source: Courtesy of ESAB Welding & Cutting Products

Figure 1.28 Portable water cooler for GMA welding equipment
Source: Courtesy of Lincoln Electric Company

Typical GMA welding machines produce a DC welding current ranging from 40 amperes to 600 amperes with 10 volts to 40 volts, depending upon the machine. In the past, some GMA processes used AC welding current, but DCRP is used almost exclusively now. Typical power supplies are shown in Figure 1.29.

Because many GMAW power supplies are used in automation and require long periods of continuous use, it is not unusual for GMA welding machines to have a 100% duty cycle. This allows the machine to be run continuously at its highest-rated output without damage.

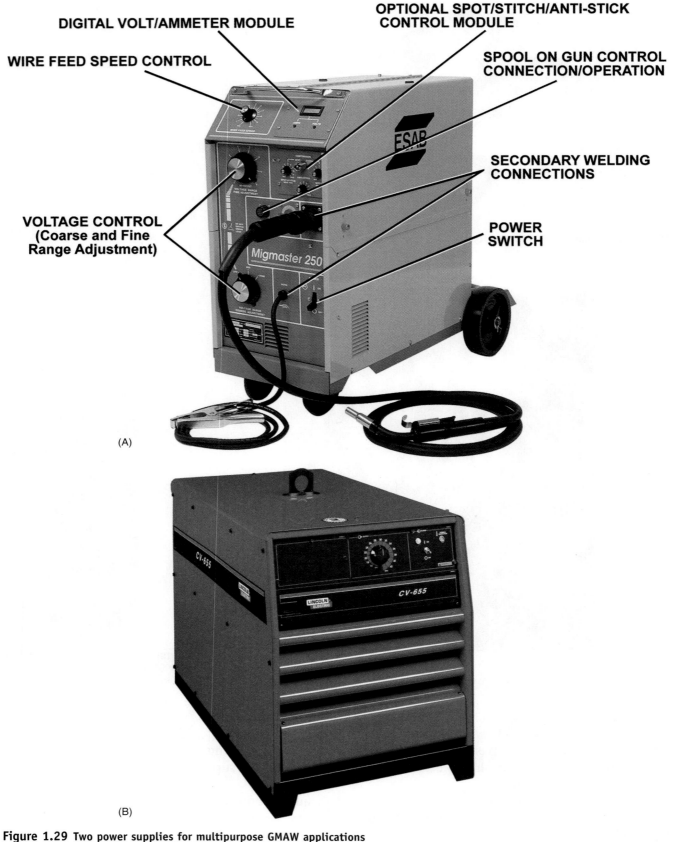

Figure 1.29 Two power supplies for multipurpose GMAW applications
(A) An expensive 200-ampere constant-voltage power supply, and (B) a 650-ampere constant-voltage and constant-current power supply.
Source: (A) Courtesy of ESAB Welding & Cutting Products (B) Courtesy of Lincoln Electric Company

Electrode (Wire) Feed Unit

The purpose of the electrode feeder is to provide a steady and reliable supply of wire to the weld. Slight changes in the rate at which the wire is fed have distinct effects on the weld.

The motor used in a feed unit is usually a DC type that can be continuously adjusted over the desired range. Figure 1.30 and Figure 1.31 show typical wire-feed units and accessories.

Push-type Feed System

The wire rollers are clamped securely against the wire to provide the necessary friction to push the wire through the conduit to the gun. The pressure applied on the wire can be adjusted. A groove is provided in the roller to aid in alignment and to lessen the chance of slippage. Most manufacturers provide rollers with smooth or knurled U-shaped or V-shaped grooves, Figure 1.32. Knurling (a series of ridges cut into the groove) helps grip larger-diameter wires so that they can be pushed along more easily. Soft wires, such as aluminum, are easy to damage if knurled rollers are used. Soft wires are best used with U-grooved rollers. Even V-grooved rollers can distort the surface of the wire, causing problems. V-grooved rollers are best suited for hard wires, such as mild steel and stainless steel. It is also important to use the correct-size grooves in the rollers.

Variations of the push-type electrode wire feeder include the pull type and push-pull type. The difference is in the size and location of the drive rollers. In the push-type system, the electrode must have enough strength to be pushed through the conduit without kinking. Mild steel and stainless steel can be readily pushed 15 to 20 ft (4 to 6 m), but aluminum is much harder to push more than 10 ft (3 m).

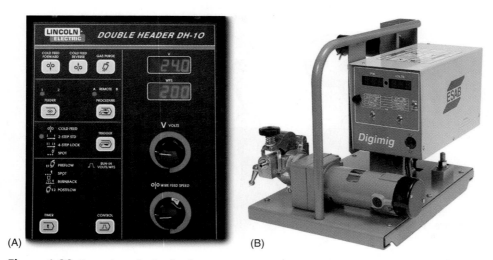

(A) (B)

Figure 1.30 Examples of wire feeders
(A) A 90-ampere power supply and wire feeder for welding sheet steel with carbon dioxide shielding. (B) Modern wire feeder with digital preset and readout of wire-feed speed and closed-loop control.
Source: (A) Courtesy of Lincoln Electric Company (B) Courtesy of ESAB Welding & Cutting Products

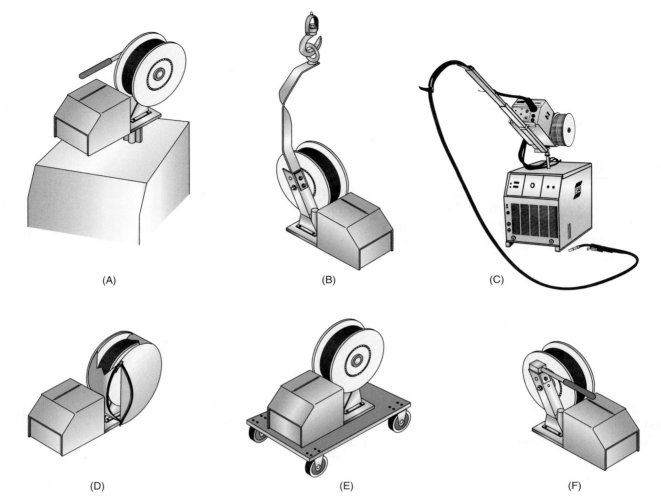

(A)   (B)   (C)

(D)   (E)   (F)

Figure 1.31 A variety of accessories are available for most electrode feed systems.
(A) Swivel post, (B) boom hanging bracket, (C) counterbalance mini-boom, (D) spool cover, (E) wire feeder wheel cart, and (F) carrying handle.

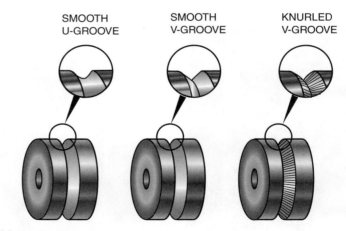

SMOOTH   SMOOTH   KNURLED
U-GROOVE   V-GROOVE   V-GROOVE

Figure 1.32 Feed rollers

Pull-type Feed System

In pull-type systems, a smaller but higher-speed motor is located in the gun to pull the wire through the conduit. Using this system, it is possible to move even soft wire over great distances. The disadvantages are that the gun is heavier and more difficult to use, rethreading the wire takes more time, and the operating life of the motor is shorter.

Push-pull-type Feed System

Push-pull-type feed systems use a synchronized system with feed motors located at both ends of the electrode conduit, Figure 1.33. This system can be used to move any type of wire over long distances by periodically installing a feed roller into the electrode conduit. Compared to the pull-type system, the advantages of this system include the ability to move wire over longer distances, faster rethreading, and increased motor life due to the reduced load. A disadvantage is that the system is more expensive.

Linear Electrode Feed System

Linear electrode feed systems use a different method to move the wire and change the feed speed. Standard systems use rollers that pinch the wire between the rollers. A system of gears is used between the motor and rollers to provide roller speed within the desired range. The linear feed system does not have gears or conventional-type rollers.

The linear feed system uses a small motor with a hollow armature shaft through which the wire is fed. The rollers are attached so that they move around the wire. Changing the roller pitch (angle) changes the speed at which the wire is moved without changing the motor speed.

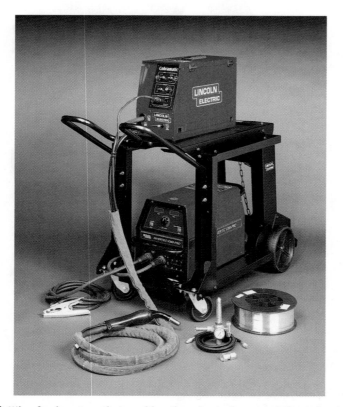

Figure 1.33 Wire-feed system that enables the wire to be moved through a longer cable
Source: Courtesy of Lincoln Electric Company

This system works in the same way that changing the pitch on a screw, either coarse threads or fine threads, affects the rate that the screw will move through a spinning nut.

The advantage of a linear system is that the bulky system of gears is eliminated, thus reducing weight, size, and wasted power. The motor operates at a constant high speed where it is more efficient. The reduced size allows the system to be housed in the gun or within an enclosure in the cable. Several linear wire feeders can be synchronized to provide an extended operating range. The disadvantage of a linear system is that the wire may become twisted as it is moved through the feeder.

Spool Gun

A spool gun is a compact, self-contained system consisting of a small drive system and a wire supply, Figure 1.34A. This system allows the welder to move freely around a job with only a power lead and shielding gas hose to manage. The major control system is usually mounted on the welder. The feed rollers and motor are found in the gun just behind the

(A)

(B)

Figure 1.34 Feeder/guns for GMA welding
Source: Courtesy of ESAB Welding & Cutting Products

nozzle and contact tube, Figure 1.34B. Because of the short distance the wire must be moved, very soft wires (aluminum) can be used. A small spool of welding wire is located just behind the feed rollers. The small spools of wire required in these guns are often very expensive. Although the guns are small, they feel heavy when being used.

Electrode Conduit

The electrode conduit or liner guides the welding wire from the feed rollers to the gun. It may be encased in a lead that contains the shielding gas.

Power cable and gun switch circuit wires are contained in a conduit that is made of a tightly wound coil having the needed flexibility and strength. The steel conduit may have a nylon or Teflon liner to protect soft, easily scratched metals, such as aluminum, as they are fed.

If the conduit is not an integral part of the lead, it must be firmly attached to both ends of the lead. Failure to attach the conduit can result in misalignment, which causes additional drag or makes the wire jam completely. If the conduit does not extend through the lead casing to make a connection, it can be drawn out by tightly coiling the lead, Figure 1.35. Coiling will force the conduit out so that it can be connected. If the conduit is too long for the lead, it should be cut off and filed smooth. Too long a lead will bend and twist inside the conduit, which may cause feed problems.

Welding Gun

The welding gun attaches to the end of the power cable, electrode conduit, and shielding gas hose, Figure 1.36. It is used by the welder to produce the weld. A trigger switch is used to start and stop the weld cycle. The gun also has a contact tube, which is used to transfer welding current to the electrode moving through the gun, and a gas nozzle, which directs the shielding gas onto the weld, Figure 1.37.

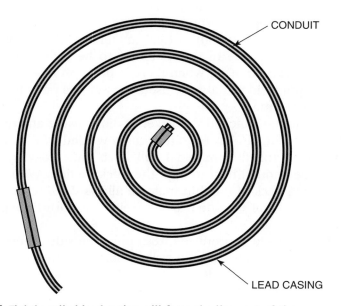

CONDUIT

LEAD CASING

Figure 1.35 Tightly coiled lead casing will force the liner out of the gun

Figure 1.36 A typical GMA welding gun
Guns like this are used for most welding processes with a heat shield attached to protect the welder's gloved hand from intense heat generated when welding with high amperages.
Source: Courtesy of Tweco, a Thermadyne Company

SPOT WELDING

GMA can be used to make high-quality arc spot welds. Welds can be made using standard or specialized equipment, Figure 1.38. The arc spot weld produced by GMAW differs from electric resistance spot welding. The GMAW spot weld starts on one surface of one member and burns through to the other member, Figure 1.39. Fusion between the members occurs, and a small nugget is left on the metal surface.

GMA spot welding has some advantages such as the following: (1) welds can be made in thin-to-thick materials; (2) the weld can be made when only one side of the materials to be welded is accessible; and (3) the weld can be made when there is paint on the interfacing surfaces. The arc spot weld can also be used to assemble parts for welding to be done at a later time.

Thin metal can be attached to thicker sections using an arc spot weld. If a thin-to-thick butt, lap, or tee joint is to be welded with complete joint penetration, often the thin material will burn back, leaving a hole, or there will not be enough heat to melt the thick section. With an arc spot weld, the burning back of the thin material allows the thicker metal to be melted. As more metal is added to the weld, the burn-through is filled, Figure 1.39.

The GMA spot weld is produced from only one side. Therefore, it can be used on awkward shapes and in cases where the other side of the surface being welded should not be damaged. This makes it an excellent process for auto body repair. In addition, because the metals are melted

Caution

Safety glasses and/or flash glasses must be worn to protect the eyes from flying sparks.

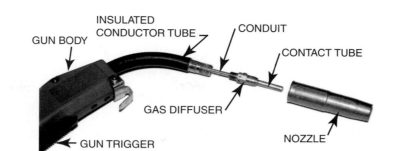

GUN BODY

INSULATED
CONDUCTOR TUBE

CONDUIT

CONTACT TUBE

GAS DIFFUSER

NOZZLE

GUN TRIGGER

(A)

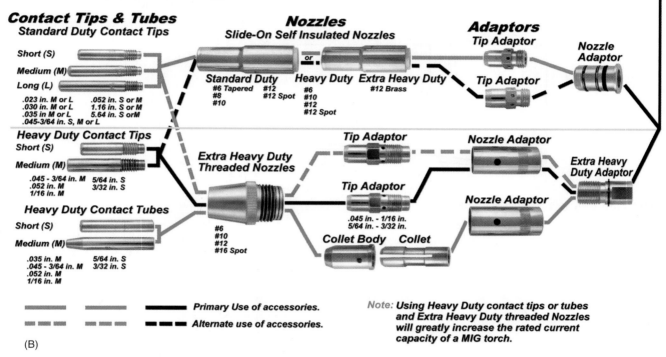

Accessories Selection Guide

**To select correct accessories choose tip based on wire,
follow chart to determine nozzle adaptor.**

200 to 500 AMP Mig Guns

Contact Tips & Tubes
Standard Duty Contact Tips

Short (S)

Medium (M)

Long (L)

.023 in. M or L   .052 in. S or M
.030 in. M or L   1.16 in. S or M
.035 in M or L   5.64 in. S orM
.045-3/64 in. S, M or L

Nozzles
Slide-On Self Insulated Nozzles

or

Standard Duty   **Heavy Duty**   **Extra Heavy Duty**
#6 Tapered   #12   #6   #12 Brass
#8   #12 Spot #10
#10   #12
                       #12 Spot

Adaptors
Tip Adaptor

Nozzle Adaptor

Tip Adaptor

Heavy Duty Contact Tips

Short (S)

Medium (M)

.045 - 3/64 in. M   5/64 in. S
.052 in. M   3/32 in. S
1/16 in. M

**Extra Heavy Duty
Threaded Nozzles**

Tip Adaptor

Nozzle Adaptor

**Extra Heavy
Duty Adaptor**

Tip Adaptor

Nozzle Adaptor

Heavy Duty Contact Tubes

Short (S)

Medium (M)

.035 in. M   5/64 in. S
.045 - 3/64 in. M   3/32 in. S
.052 in. M
1/16 in. M

#6
#10
#12
#16 Spot

.045 in. - 1/16 in.
5/64 in. - 3/32 in.

Collet Body   **Collet**

 **Primary Use of accessories.**

Alternate use of accessories.

**Note: Using Heavy Duty contact tips or tubes
and Extra Heavy Duty threaded Nozzles
will greatly increase the rated current
capacity of a MIG torch.**

(B)

Figure 1.37 GMA welding gun parts
(A) Typical replaceable parts of a GMA welding gun. (B) Accessories and parts selection guide for a GMA welding gun.
Source: Courtesy of ESAB Welding & Cutting Products

and the molten weld pool is agitated, thin films of paint between the members being joined need not be removed. This is an added benefit for auto body repair work.

Specially designed nozzles provide flash protection, part alignment, and arc alignment, Figure 1.40. As a result, for some small jobs it is possible to perform the weld with only safety glasses. The optional control timer provides weld time and burn-back time. To make a weld, the amperage, voltage, and length of welding time must be set correctly. The burn-back time is a short period at the end of the weld when the wire feed stops but the current does not. This allows the wire to be burned back so it does not stick in the weld, Figure 1.39D.

Caution

It is not advisable for any spot welding work requiring more than just a few spot welds to be done without full welder's safety gear. Prolonged exposure to the reflected ultraviolet light will cause skin burns.

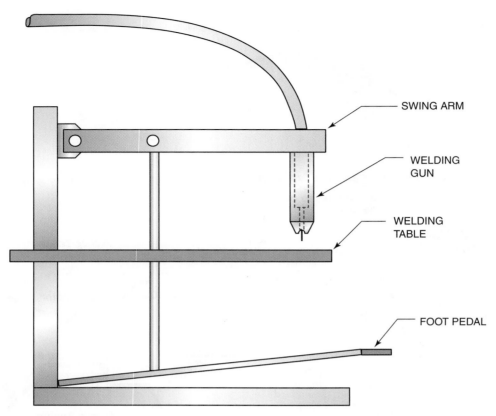

Figure 1.38 GMA spot welding machine

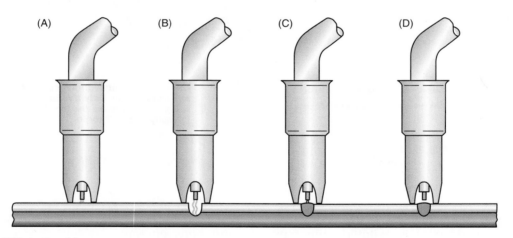

Figure 1.39 A GMA spot weld
(A) The arc starts, (B) a hole is burned through the first plate, (C) the hole is filled with weld metal, and (D) the wire feed stops and the arc burns the electrode back.

Module 2
Key Indicator 1

Module 5
Key Indicator 1

FACE, EYE, AND EAR PROTECTION

Face and Eye Protection

Eye protection must be worn in the shop at all times. Eye protection can be safety glasses with side shields, Figure 1.41, goggles, or a full face shield. To give better protection when working in brightly lit areas or outdoors, some welders wear flash glasses, which are special, lightly tinted, safety glasses. These safety glasses provide protection from both flying debris and reflected light.

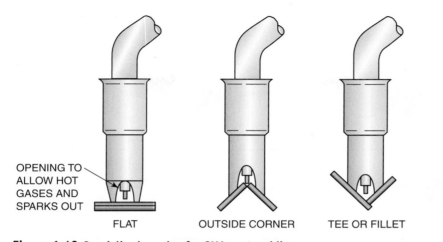

OPENING TO ALLOW HOT GASES AND SPARKS OUT

FLAT   OUTSIDE CORNER   TEE OR FILLET

SIDE SHIELDS

Figure 1.40 Specialized nozzles for GMA spot welding

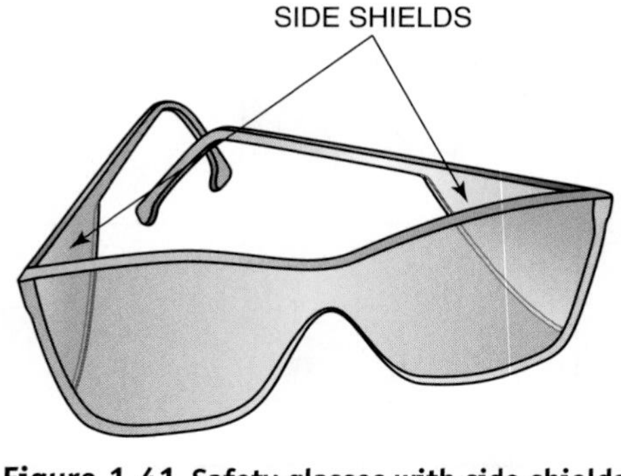

Figure 1.41 Safety glasses with side shields

Suitable eye protection is important because eye damage caused by excessive exposure to arc light is not noticed. Welding light damage often occurs without warning, like a sunburn's effect that is felt the following day. Therefore, welders must take appropriate precautions in selecting filters or goggles that are suitable for the process being used, Figure 1.42. Selecting the correct shade lens is also important, because both extremes of too light or too dark can cause eye strain. New welders often select too dark a lens, assuming it will give them better protection, but this results in eye strain in the same manner as if they were trying to read in a poorly lit room. In reality, any approved arc welding lenses will filter out the harmful ultraviolet light. Select a lens that lets you see comfortably. At the very least, the welder's eyes must not be strained by excessive glare from the arc.

Ultraviolet light can burn the eye in two ways. This light can injure either the white of the eye or the retina, which is the back of the eye, Figure 1.43. Burns on the retina are not painful but may cause some loss of eyesight. The whites of the eyes are very sensitive, and burns are very painful. The eyes are easily infected because, as with any burn, many cells are killed. These dead cells in the moist environment of the eyes will promote the growth of bacteria that cause infection. When the eye is burned, it feels as though there is something in the eye. Without a professional examination, however, it is impossible to tell if there is something in the eye. Because there may actually be something in the eye and because of the high risk of infection, home remedies or other medicines should never be used for eye burns. Anytime you receive an eye injury you should see a doctor.

Welding Helmets

Even with quality **welding helmets**, like those shown in Figure 1.44, the welder must check for potential problems that may occur from accidents or daily use. Small, undetectable leaks of ultraviolet light in an arc welding helmet can cause a welder's eyes to itch or feel sore after a day of welding. To prevent these leaks, make sure the lens gasket is installed correctly, Figure 1.45. The outer and inner clear lenses must be plastic. As shown in Figure 1.46, a lens can be checked for cracks by twisting it between your fingers. Worn or cracked spots on a helmet must be repaired.

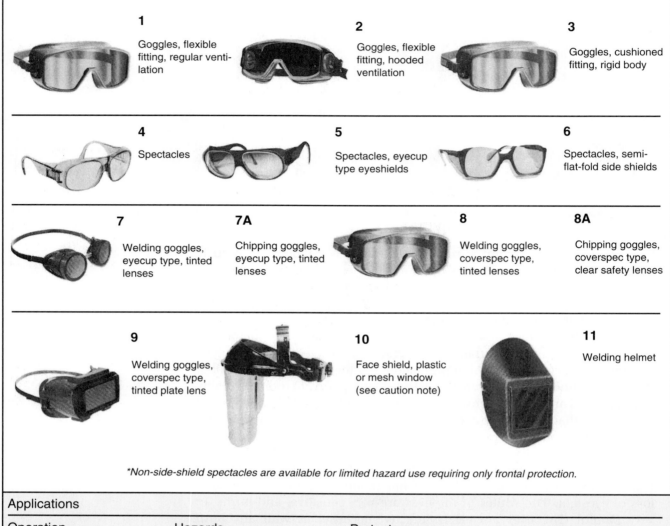

1 Goggles, flexible fitting, regular ventilation

2 Goggles, flexible fitting, hooded ventilation

3 Goggles, cushioned fitting, rigid body

4 Spectacles

5 Spectacles, eyecup type eyeshields

6 Spectacles, semi-flat-fold side shields

7 Welding goggles, eyecup type, tinted lenses

7A Chipping goggles, eyecup type, tinted lenses

8 Welding goggles, coverspec type, tinted lenses

8A Chipping goggles, coverspec type, clear safety lenses

9 Welding goggles, coverspec type, tinted plate lens

10 Face shield, plastic or mesh window (see caution note)

11 Welding helmet

Non-side-shield spectacles are available for limited hazard use requiring only frontal protection.

Applications

Operation	Hazards	Protectors
Acetylene-burning Acetylene-cutting Acetylene-welding	Sparks, harmful rays, molten metal, flying particles	7,8,9
Chemical handling	Splash, acid burns, fumes	2 (for severe exposure add 10)
Chipping	Flying particles	1,2,4,5,6,7A,8A
Electric (arc) welding	Sparks, intense rays, molten metal	11 (in combination with 4,5,6 in tinted lenses advisable)
Furnace operations	Glare, heat, molten metal	7,8,9 (for severe exposure add 10)
Grinding—light	Flying particles	1,3,5,6 (for severe exposure add 10)
Grinding—heavy	Flying particles	1,3,7A,8A (for severe exposure add 10)
Laboratory	Chemical splash, glass breakage	2 (10 when in combination with 5,6)
Machining	Flying particles	1,3,5,6 (for severe exposure add 10)
Molten metals	Heat, glare, sparks, splash	7,8 (10 in combination with 5,6 in tinted lenses)
Spot welding	Flying particles, sparks	1,3,4,5,6 (tinted lenses advisable, for severe exposure add 10)

CAUTION:
Face shields alone do not provide adequate protection. Plastic lenses are advised for protection against molten metal splash.
Contact lenses, of themselves, do not provide eye protection in the industrial sense and shall not be worn in a hazardous environment without appropriate covering safety eyewear.

Figure 1.42 Huntsman selector chart
Source: Courtesy of Kedman Co., Huntsman Product Division

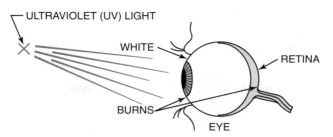
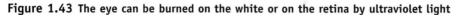

Figure 1.43 The eye can be burned on the white or on the retina by ultraviolet light

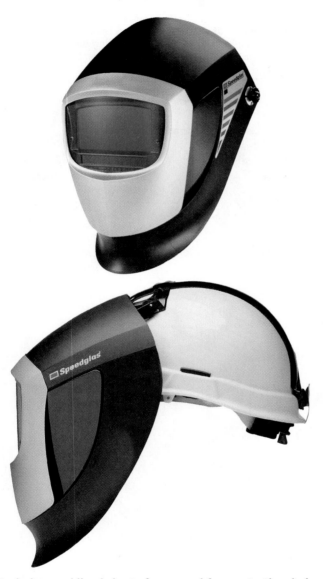

Figure 1.44 Typical arc welding helmets for eye and face protection during welding
Source: Courtesy of Hornell, Inc.

Safety Glasses

Safety glasses with side shields are adequate for general use, but if heavy grinding, chipping, or overhead work is being done, goggles or a full face shield should be worn in addition to safety glasses, Figure 1.47. Safety glasses are best for general protection. They must be worn under an arc welding helmet at all times.

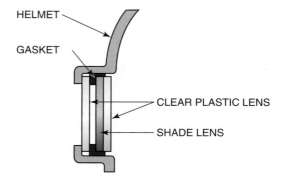

HELMET

GASKET

CLEAR PLASTIC LENS

SHADE LENS

Figure 1.45 Correct placement of the gasket around the shade lens
Correct gasket placement around the shade lens of a welding helmet is important because it can stop ultraviolet light from bouncing around the lens assembly.

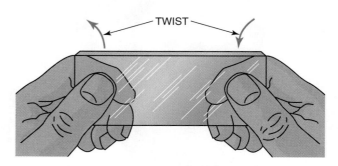

TWIST

Figure 1.46 To check the shade lens for possible cracks, gently twist it

Figure 1.47 Full face shield

Ear Protection

The welding environment can be very noisy. The sound level is at times high enough to cause pain and some loss of hearing if the welder's ears are unprotected. Hot sparks can also drop into an open ear, causing severe burns.

Ear protection is available in several forms. One form of protection is earmuffs that cover the outer ear completely, Figure 1.48. Another form

Figure 1.48 Earmuffs
Earmuffs provide complete ear protection and can be worn under a welding helmet.
Source: Courtesy of Mine Safety Appliances Company

Figure 1.49 Earplugs
Earplugs are used as protection from noise only.
Source: Courtesy of Mine Safety Appliances Company

of protection is earplugs that fit into the ear canal, Figure 1.49. Both of these protect a person's hearing, but only the earmuffs protect the outer ear from burns.

GENERAL WORK CLOTHING

Special protective clothing cannot be worn at all times. It is, therefore, important to choose general work clothing that will minimize the possibility of getting burned because of the high temperature and amount of hot sparks, metal, and slag produced during welding, cutting, or brazing.

Caution

Damage to your hearing caused by high sound levels may not be detected until later in life, and the resulting loss in hearing is permanent. Your hearing will not improve with time, and each exposure to high levels of sound will further damage your hearing.

Work clothing must also stop ultraviolet light from passing through it. This is accomplished if the material chosen is a dark color, thick, and tightly woven. The best choice is 100% wool but it is difficult to find. Another good choice is 100% cotton clothing, the most popular fabric used.

You must avoid wearing synthetic materials including nylon, rayon, and polyester. They can easily melt or catch fire. Some synthetics produce a hot, sticky residue that can make burns more severe. Others may produce poisonous gases.

The following are some guidelines for selecting work clothing:

- Shirts must be long-sleeved to protect the arms, have a high-buttoned collar to protect the neck, Figure 1.50, be long enough to tuck into the pants to protect the waist, and have flaps on the pockets to keep sparks out (or have no pockets).
- Pants must have legs long enough to cover the tops of the boots and must be without cuffs that would catch sparks.
- Boots must have high tops to keep out sparks, have steel toes to prevent crushed toes, Figure 1.51, and have smooth tops to prevent sparks from being trapped in seams.
- Caps should be thick enough to prevent sparks from burning the top of a welder's head.

Figure 1.50 Neck protection
The top button of the shirt worn by the welder should always be buttoned in order to avoid severe neck burns.
Source: Courtesy of Larry Jeffus

STEEL

Figure 1.51 Safety boots with steel toes are required by many welding shops

All clothing must be free of frayed edges and holes. The clothing must be relatively tight-fitting in order to prevent excessive folds or wrinkles that might trap sparks.

Some welding clothes have pockets on the inside to prevent the pockets from collecting sparks. However, it is not safe to carry butane lighters or matches in these or any pockets while welding. Lighters and matches can easily catch on fire or explode if they are subjected to the heat and sparks of welding.

SPECIAL PROTECTIVE CLOTHING

General work clothing is worn by each person in the shop. In addition to this clothing, extra protection is needed for each person who is in direct contact with hot materials. Leather is often the best material to use, as it is lightweight, is flexible, resists burning, and is readily available. Synthetic insulating materials are also available. Ready-to-wear leather protection includes capes, jackets, aprons, sleeves, gloves, caps, pants, knee pads, and spats, among other items.

Hand Protection

All-leather, gauntlet-type gloves should be worn when doing any welding, Figure 1.52. Gauntlet gloves that have a cloth liner for insulation are best for hot work. Noninsulated gloves will give greater flexibility for fine work. Some leather gloves are available with a canvas gauntlet top, which should be used for light work only.

When a great deal of manual dexterity is required for gas tungsten arc welding, brazing, soldering, oxyfuel gas welding, and other delicate processes, soft leather gloves may be used, Figure 1.53. All-cotton gloves are sometimes used when doing very light welding.

> ## Caution
>
> There is no safe place to carry butane lighters or matches while welding or cutting. They can catch fire or explode if subjected to welding heat or sparks. Butane lighters may explode with the force of a quarter of a stick of dynamite. Matches can erupt into a ball of fire. Both butane lighters and matches must always be removed from the welder's pockets and placed a safe distance away before any work is started.

Figure 1.52 All-leather, gauntlet-type welding gloves
Source: Courtesy of Larry Jeffus

Figure 1.53 Soft leather gloves
For welding that requires a great deal of manual dexterity, soft leather gloves can be worn.
Source: Courtesy of Larry Jeffus

Body Protection

Full leather jackets and capes will protect a welder's shoulders, arms, and chest, Figure 1.54. A jacket, unlike the cape, protects a welder's back and complete chest. A cape is open and much cooler but offers less protection. The cape can be used with a bib apron to provide some additional protection while leaving the back cooler. Either the full jacket or the cape with a bib apron should be worn for any out-of-position work.

Waist and Lap Protection

Bib aprons or full aprons will protect a welder's lap. Welders will especially need to protect their laps if they squat or sit while working and when they bend over or lean against a table.

Arm Protection

For some vertical welding, a full or half sleeve can protect a person's arm, Figure 1.55. The sleeves work best if the work level is not above the welder's chest. Work levels higher than this usually require a jacket or cape to keep sparks off the welder's shoulders.

Leg and Foot Protection

When heavy cutting or welding is being done and a large number of sparks are falling, leather pants and spats should be used to protect the welder's legs and feet. If the weather is hot and full leather pants are uncomfortable, leather aprons with leggings are available. Leggings can be strapped to the legs, leaving the back open. Spats will prevent sparks from burning through the front of lace-up boots.

Figure 1.54 Full leather jacket
Source: Courtesy of Larry Jeffus

Figure 1.55 Full leather sleeve
Source: Courtesy of Larry Jeffus

SUMMARY

The keys to producing quality GMA welds are equipment, setup, and adjustments. Once you have mastered these skills the only remaining obstacle to your producing consistent, uniform, high-quality welds is your ability to follow, or track, the joint consistently. Some welders find that lightly dragging their glove along the metal surface or edge of the fabrication can aid them in controlling the weld consistency. One of the advantages of the GMA welding process is its ability to produce long, uninterrupted welds. However, this often leads to welder fatigue. Finding a comfortable welding position that you can maintain for several minutes at a time will both improve your weld quality and reduce your fatigue.

Selecting the proper method of metal transfer—short arc, globular, or spray transfer—is normally done by the welding shop foreman or supervisor. He or she makes these selections based on the material being welded, the welding position, and other factors, including welding procedure specifications and applicable codes. A welder must be proficient with each of the various methods of metal transfer. It is therefore important that you spend time practicing and developing your skills with each of these processes.

REVIEW

1. Why is usage of the term *GMAW* preferable to *MIG* for gas metal arc welding?
2. Using Table 1.1, answer the following:
   a. What maintains the arc in machine welding?
   b. What feeds the filler metal in manual welding?
   c. What provides the joint travel in automatic welding?
   d. What provides the joint guidance in semiautomatic welding?
3. What factors have led to increased usage of the GMAW process?
4. In what form is metal transferred across the arc in the axial spray metal transfer method of GMA welding?
5. What three conditions are required for the spray transfer process to occur?
6. Using Table 1.3, answer the following:
   a. What should the wire-feed speed and voltage ranges be to weld 1/8-in. metal with 0.035-in. wire using argon shielding gas?
   b. What should the amperage and voltage ranges be when using 98% Ar + 2% O_2 to weld 1/4-in. metal with 0.045-in. wire?
7. What ranges does the pulsed-arc metal transfer shift between?
8. How do frequency, amplitude, and width of the pulses affect the GMA pulse welding process?
9. How have electronics helped the pulsed-arc process?
10. Why is helium added to argon when making some spray or pulsed-spray transfer welds?
11. Why does DCEP help with welds on metals such as aluminum?
12. Why is CO_2 added to argon when making GMA spray transfer welds?
13. Why should CO_2 not be used to weld stainless steel?
14. How is the metal transferred from the electrode to the plate during the GMAW-S process?
15. Using Figure 1.12, what should be the approximate voltage at 175 amps at 200 in./min when using 0.035-in. ER70S-6 electrode wire?
16. Using Table 1.8, what would the amperage be for 0.035-in. (0.9-mm) wire at 200 in./min (5 m/min)?
17. Using Table 1.9, what shielding gas should be used for welding on copper?
18. What may happen if the GMA welding electrode is allowed to strike the base metal outside the molten weld pool?
19. What effect does shortening the electrode extension have on weld penetration?
20. Describe the weld produced by a backhand welding angle.
21. Describe the weld produced by a forehand welding angle.
22. What components make up a GMA welding system?
23. Why must GMA welders have a 100% duty cycle?
24. What can happen if rollers of the wrong shape are used on aluminum wire?
25. Where is the drive motor located in a pull-type wire-feed system?
26. How is the wire-feed speed changed with a linear feed system?
27. What type of liner should be used for aluminum wire?
28. What parts of a typical GMA welding gun can be replaced?
29. Describe the spot welding process using a GMA welder.

CHAPTER 2

Gas Metal Arc Welding

KEY TERMS

bird-nesting	contact tube	spool drag
cast	feed rollers	wire-feed speed
conduit liner	flow rate	

AWS SENSE EG2.0

Key Indicators Addressed in this Chapter:

Module 1: Occupational Orientation

Key Indicator 1: Prepares time or job cards, reports or records
Key Indicator 2: Performs housekeeping duties
Key Indicator 3: Follows verbal instructions to complete work assignments
Key Indicator 4: Follows written details to complete work assignments

Module 2: Safety and Health of Welders

Key Indicator 1: Demonstrates proper use and inspection of personal protection equipment (PPE)
Key Indicator 2: Demonstrates proper work area operation

Key Indicator 3: Demonstrates proper use and inspection of ventilation equipment

Key Indicator 4: Demonstrates proper hot zone operation

Key Indicator 7: Demonstrates proper inspection and operation of equipment for each welding or thermal cutting process used (This is best done as part of the process module/unit for each of the required welding and thermal cutting processes.)

Module 3: Drawing and Welding Symbol Interpretation

Key Indicator 1: Interprets basic elements of a drawing or sketch

Key Indicator 2: Interprets welding symbol information

Key Indicator 3: Fabricates parts from a drawing or sketch

Module 5: Gas Metal Arc Welding [GMAW-S, GMAW (spray)]

Key Indicator 1: Performs safety inspections of GMAW equipment and accessories

Key Indicator 2: Makes minor external repairs to GMAW equipment and accessories

Short-circuit Transfer

Key Indicator 3: Sets up for GMAW-S operations on carbon steel

Key Indicator 4: Operates GMAW-S equipment on carbon steel

Key Indicator 5: Makes GMAW-S fillet welds, in all positions, on carbon steel

Key Indicator 6: Makes GMAW-S groove welds, in all positions, on carbon steel

Key Indicator 7: Passes GMAW-S welder performance qualification test (workmanship sample) on carbon steel

Spray Transfer

Key Indicator 8: Sets up for GMAW (spray) operations on carbon steel

Key Indicator 9: Operates GMAW (spray) equipment on carbon steel

Key Indicator 10: Makes GMAW (spray) fillet welds, in the 1F and 2F positions, on carbon steel

Key Indicator 11: Makes GMAW (spray) groove welds, in the 1G position, on carbon steel

Key Indicator 12: Passes GMAW (spray) welder performance qualification test (workmanship sample) on carbon steel

Module 9: Welding Inspection and Testing Principles

Key Indicator 1: Examines cut surfaces and edges of prepared base metal parts.

Key Indicator 2: Examines tacks, root passes, intermediate layers, and completed welds

INTRODUCTION

Performing a satisfactory GMA weld requires more than just manipulative skill. The setup, voltage, amperage, electrode extension, and welding angle, as well as other factors, can dramatically affect the weld produced. The very best welding conditions are those that will allow a welder to produce the largest quantity of successful welds in the shortest period of time with the highest productivity. Because these are semiautomatic or automatic processes, increased productivity may require only that the welder increase the travel speed and current. This does not mean that the welder will work harder but, rather, that the welder will work more productively, resulting in a greater cost efficiency.

The more cost efficient welders can be, the more competitive they and their companies become. This can make the difference between being awarded a job or losing work.

SETUP

The same equipment may be used for semiautomatic GMAW, flux cored arc welding (FCAW), and submerged arc welding (SAW). Often, FCAW and SAW equipment have a higher amperage range. In addition, equipment for FCAW and SAW is more likely to be automated than that for GMAW. However, GMA welding equipment can easily be automated.

The basic GMAW installation consists of the following: welding gun, gun switch circuit, electrode conduit or liner, welding contractor control, electrode feed unit, electrode supply, power source, shielding gas supply, shielding gas flowmeter regulator, shielding gas hoses, and both power and work cables. Typical water-cooled and air-cooled guns are shown in Figure 2.1. The equipment setup in this chapter is similar to equipment built by other manufacturers, which means that any skills developed can be transferred easily to other equipment.

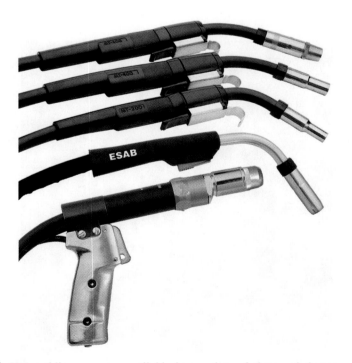

Figure 2.1 GMA welding guns are available in a variety of sizes and shapes
Source: Courtesy of ESAB Welding & Cutting Products

PRACTICE 2-1

GMAW Equipment Setup

For this practice, you will need a GMAW power source, a welding gun, an electrode feed unit, an electrode supply, a shielding gas supply, a shielding gas flowmeter regulator, electrode conduit, power and work leads, shielding gas hoses, assorted hand tools, spare parts, and any other required materials. In this practice, you will properly set up a GMA welding installation.

If the shielding gas supply is a cylinder, it must be chained securely in place before the valve protection cap is removed, Figure 2.2. Standing to one side of the cylinder and making sure no bystanders are in line with the valve, quickly crack the valve to blow out any dirt in the valve before the flowmeter regulator is attached, Figure 2.3. With the flowmeter regu-

Figure 2.2 Make sure the gas cylinder is chained securely in place before removing the safety cap
Source: Courtesy of Larry Jeffus

Figure 2.3 Attaching the flowmeter regulator
Be sure the tube is vertical.
Source: Courtesy of Larry Jeffus

lator attached securely to the cylinder valve, attach the correct hose from the flowmeter to the "gas-in" connection on the electrode feed unit or machine.

Install the reel of electrode (welding wire) on the holder and secure it, Figure 2.4. Check the feed roller size to ensure that it matches the wire size, Figure 2.5. The conduit liner size should be checked to be sure that it is compatible with the wire size. Connect the conduit to the feed unit. The conduit or an extension should be aligned with the groove in the roller and set as close to the roller as possible without touching, Figure 2.6. Misalignment at this point can contribute to a bird's nest, Figure 2.7. **Bird-nesting** of the electrode wire results when the feed roller pushes the wire into a tangled ball, like a bird's nest, because the wire would not go through the outfeed side conduit.

Figure 2.4 Wire label
When installing the spool of wire, check the label to be sure that the wire is the correct type and size.
Source: Courtesy of Larry Jeffus

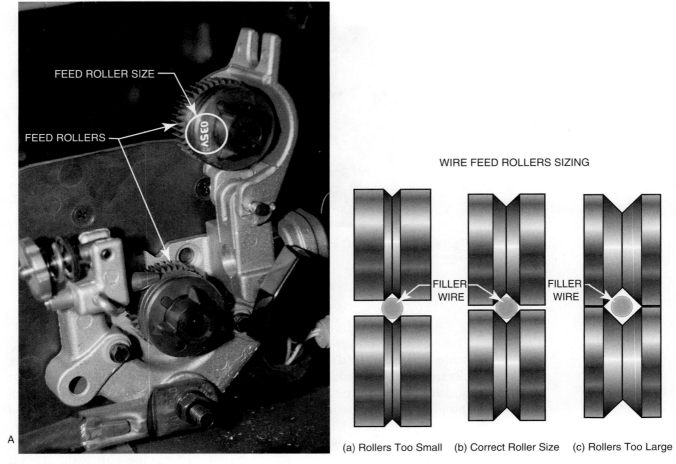

Figure 2.5 Checking feed roller size
(A) Check to be certain that the feed rollers are the correct size for the wire being used. (B) If the wirefeed rollers are too small, the welding wire could be damaged. If the wirefeed rollers are too large, the rollers will not grip the wire.
Source: (A) Courtesy of Larry Jeffus

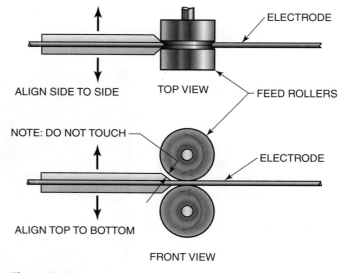

Figure 2.6 Feed

Figure 2.7 "Bird's nest" in the filler wire at the feed rollers
Source: Courtesy of Larry Jeffus

   Be sure the power is off before attaching the welding cables. The electrode and work leads should be attached to the proper terminals. The electrode lead should be attached to the terminal marked electrode or positive (+). If necessary, it is also attached to the power cable part of the gun lead. The work lead should be attached to the terminal marked work or negative (–).

   The shielding "gas-out" side of the solenoid is then also attached to the gun lead. If a separate splice is required from the gun switch circuit to the feed unit, it should be connected at this time, Figure 2.8. Check to see that the welding contractor circuit is connected from the feed unit to the power source.

The welding gun should be securely attached to the main lead cable and conduit, Figure 2.9. There should be a gas diffuser attached to the end of the conduit liner to ensure proper alignment. A **contact tube** (tip) of the correct size to match the electrode wire size being used should be installed, Figure 2.10. A shielding gas nozzle is attached to complete the assembly.

Recheck all fittings and connections for tightness. Loose fittings can leak; loose connections can cause added resistance, reducing the welding

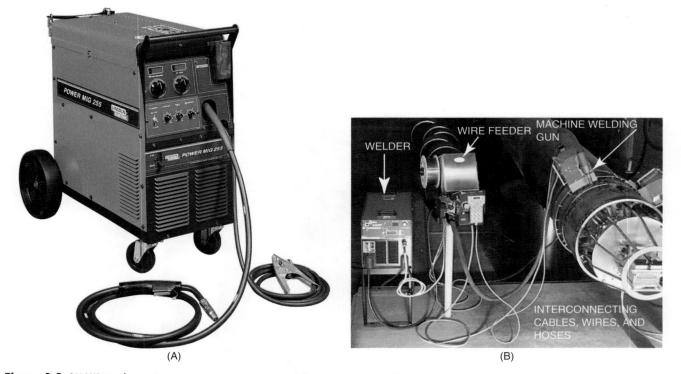

(A)   (B)

Figure 2.8 GMAW station setup
(A) Typical GMA welding machine. (B) Typical interconnecting cables and wires for a semiautomatic GMA welding station.
Source: (A) Courtesy of Lincoln Electric Company (B) Courtesy of Dynatorque

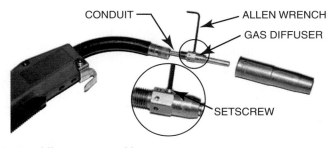

Figure 2.9 GMA welding gun assembly
Source: Courtesy of Larry Jeffus

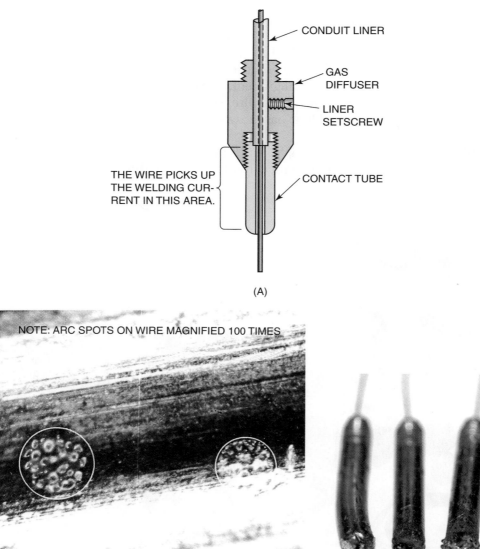

CONDUIT LINER

GAS DIFFUSER

LINER SETSCREW

THE WIRE PICKS UP THE WELDING CURRENT IN THIS AREA.

CONTACT TUBE

(A)

NOTE: ARC SPOTS ON WIRE MAGNIFIED 100 TIMES

(B)

(C)

Figure 2.10 The contact tube must be the correct size
(A) Too small a contact tube will cause the wire to stick. (B) Too large a contact tube can cause arcing to occur between the wire and tube. (C) Heat from the arcing can damage the tube.
Source: (B) Courtesy of Brett V. Hahn (C) Courtesy of Larry Jeffus

efficiency. Some manufacturers include detailed setup instructions with their equipment, Figure 2.11.

Complete a copy of the "Student Welding Report" listed in Appendix I or provided by your instructor.

Module 1
Key Indicator 1, 2, 3, 4

Module 2
Key Indicator 1, 2, 3, 4, 7

Module 5
Short-circuit Transfer
Key Indicator 3
Spray Transfer
Key Indicator 8

PRACTICE 2-2

Threading GMAW Wire

Using the GMAW machine that was properly assembled in Practice 2-1, you will turn the machine on and thread the electrode wire through the system.

Open the side cover.

With the gun trigger pressed, adjust the feed roller tension.

Remove the empty wire spool.

Check the setting guide inside the machine door.

Release upper feed roller.

Set the voltage and wire feed for the metal you are going to be welding.

Reload the wire spool with the free end unreeling from the bottom.

Attach work cable clamp to work to be welded.

Thread wire through guide between rollers and into wire cable.

Connect gas to coupling at rear of case and turn on shielding gas.

Set the polarity as DCEP from GMA welding.

ALWAYS WEAR PROPER SAFETY EQUIPMENT. Pull trigger and weld.

Turn the input switch on.

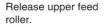

Figure 2.11 Example of manufacturer's setup instructions
Source: Courtesy of Lincoln Electric Company

Check to see that the unit is assembled correctly according to the manufacturer's specifications. Switch on the power and check the gun switch circuit by depressing the switch. The power source relays, feed relays, gas solenoid, and feed motor should all activate.

Cut the end of the electrode wire free. Hold it tightly so that it does not unwind. The wire has a natural curve that is known as its **cast**. The cast is measured by the diameter of the circle that the wire would make if it were loosely laid on a flat surface, Figure 2.12. The cast helps the wire make a good electrical contact as it passes through the contact tube. However, the cast can be a problem when threading the system. To make threading easier, straighten about 12 in. (305 mm) of the end of the wire and cut any kinks off.

Separate the wire **feed rollers** and push the wire first through the guides, Figure 2.13, then between the rollers, and finally into the **conduit liner**. Reset the rollers so there is a slight amount of compression on the wire, Figure 2.14. Set the **wire-feed speed** control to a slow speed. Hold the welding gun so that the electrode conduit and cable are as straight as possible.

With safety glasses on and the gun pointed away from the welder's face, press the gun switch, or the cold feed switch if your wire feeder is equipped with one. The cold feed switch feeds wire without delivering current to the gun. The wire should start feeding into the liner. Watch to make certain that the wire feeds smoothly and release the gun switch as soon as the end comes through the contact tube.

With the wire feed running, adjust the feed roller compression so that the wire reel can be stopped easily by a slight pressure. Too light a roller pressure will cause the wire to feed erratically. Too high a pressure can turn a minor problem into a major disaster. If the wire jams at a high roller pressure, the feed rollers keep feeding the wire, causing it to bird-nest and possibly short out. With a light pressure, the wire can stop, preventing bird-nesting. This is very important with soft wires. The other

Caution

If the wire stops feeding before it reaches the end of the contact tube, stop and check the system. If no obvious problem can be found, mark the wire with tape and remove it from the gun. It then can be held next to the system to determine the location of the problem.

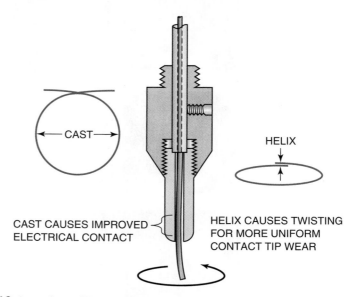

CAST

HELIX

CAST CAUSES IMPROVED ELECTRICAL CONTACT

HELIX CAUSES TWISTING FOR MORE UNIFORM CONTACT TIP WEAR

Figure 2.12 Cast of a welding wire
The cast of the welding wire causes it to rub firmly inside the contact tube for good electrical contact. The helix causes the electrode to twist inside the contact tube so that the tube is worn uniformly.

Figure 2.13 Push the wire through the guides by hand
Source: Courtesy of Larry Jeffus

Figure 2.14 Adjust the wire-feed tensioner
Source: Courtesy of Larry Jeffus

advantage of a light pressure is that the feed will stop if something like clothing or a gas hose is caught in the reel.

With the feed running, adjust the **spool drag** so that the reel stops when the feed stops. The reel should not coast to a stop because the wire can be snagged easily. Also, when the feed restarts, a jolt occurs when the slack in the wire is taken up. This jolt can be enough to momentarily stop the wire, possibly causing a discontinuity in the weld.

When the test runs are completed, the wire can either be rewound or cut off. Some wire-feed units have a retract button. This allows the feed driver to reverse and retract the wire automatically. To rewind the wire on units without this retraction feature, release the rollers and turn them backward by hand. If the machine will not allow the feed rollers to be released without upsetting the tension, you must cut the wire.

Complete a copy of the "Student Welding Report" listed in Appendix I or provided by your instructor.

GAS DENSITY AND FLOW RATES

Density is the chief determinant of how effective a gas is for arc shielding. The lower the density of a gas, the higher will be the flow rate required for equal arc protection. Flow rates, however, are not in proportion to the

Caution

Do not discard pieces of wire on the floor. They present a hazard to safe movement around the machine. In addition, a small piece of wire can work its way into a filter screen on the welding power source. If the piece of wire shorts out inside the machine, it could become charged with high voltage, which could cause injury or death. Always wind the wire tightly into a ball or cut it into short lengths before discarding it in the proper waste container.

NOTE

If you need more shielding gas coverage in a windy or drafty area, use both a larger-diameter gas nozzle and a higher gas flow rate. The larger the nozzle size, the higher the permissible flow rate without causing turbulence. Larger nozzle sizes may restrict your view of the weld. You might also consider setting up a wind barrier to protect your welding from the wind, Figure 2.16.

Module 1
Key Indicator 1, 2, 4

Module 5
Short-circuit Transfer
Key Indicator 3
Spray Transfer
Key Indicator 8

densities. Helium, with about one-tenth the density of argon, requires about twice the flow for equal protection.

The correct flow rate can be set by checking welding guides that are available from the welding equipment and filler metal manufacturers. These welding guides list the gas flow required for various nozzle sizes and welding amperage settings. Some welders feel that a higher gas flow will provide better weld coverage, but that is not always the case. High gas flow rates waste shielding gases and may lead to contamination. The contamination comes from turbulence in the gas at high flow rates. Air is drawn into the gas envelope by the venturi effect around the edge of the nozzle. Also, the air can be drawn in under the nozzle if the torch is held at too sharp an angle to the metal, Figure 2.15.

EXPERIMENT 2-1

Setting Gas Flow Rate

Using the equipment setup as described in Practice 2-1, and the threaded machine as described in Practice 2-2, you will set the shielding gas flow rate.

The exact **flow rate** required for a certain job will vary depending upon welding conditions. This experiment will help you determine how those conditions affect the flow rate. You will start by setting the shielding gas flow rate at 35 cubic feet per hour (cfh) (16 L/min).

Standing to one side, turn on the shielding gas supply valve. If the supply is a cylinder, the valve is opened all the way. With the machine power on and the welding gun switch depressed, you are ready to set

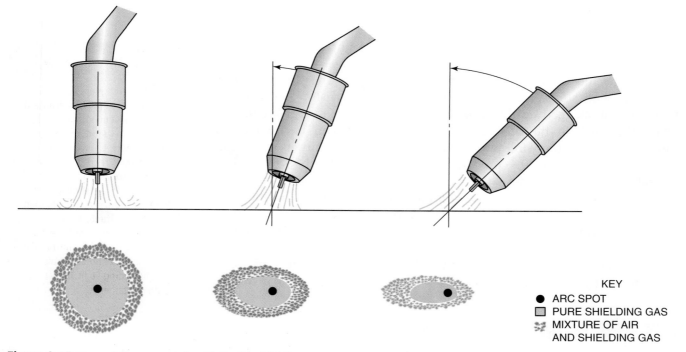

KEY
● ARC SPOT
▢ PURE SHIELDING GAS
✂ MIXTURE OF AIR
  AND SHIELDING GAS

Figure 2.15 The welding gun angle affects the shielding gas coverage for the molten weld pool.

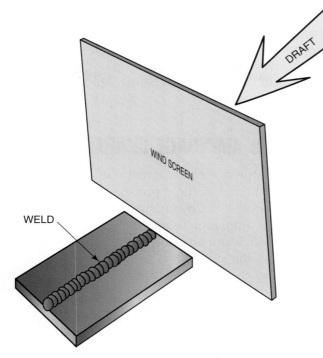

Figure 2.16 A wind screen can help prevent the shielding gas from being blown away.

the flow rate. Slowly turn in the adjusting screw and watch the float ball as it rises in a tube on a column of gas. The faster the gas flows, the higher the ball will float. A scale on the tube allows you to read the flow rate. Different scales are used with each type of gas being used. Since various gases have different densities (weights), the ball will float at varying levels even though the flow rates are the same, Figure 2.17. The line corresponding to the flow rate may be read as it compares to the top, center, or bottom of the ball, depending upon the manufacturer's instructions. There should be some marking or instruction on the tube or regulator to tell how it should be read, Figure 2.18.

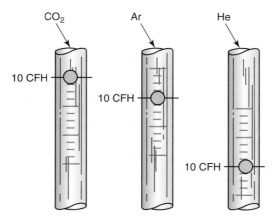

Figure 2.17 Reading gas flow rates
Each of these gases is flowing at the same cfh (L/min) rate. Because helium (He) is less dense, its indicator ball is the lowest. Be sure that you are reading the correct scale for the gas being used.

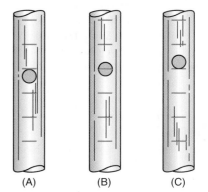

Figure 2.18 Three methods of reading a flowmeter
(A) Top of ball, (B) center of ball, and (C) bottom of ball.

Release the welding gun switch, and the gas flow should stop. Turn off the power and spray the hose fittings with a leak-detecting solution.

When you stop for more than a short period, close the shielding gas supply valve and release the hose pressure.

Complete a copy of the "Student Welding Report" listed in Appendix I or provided by your instructor.

ARC-VOLTAGE AND AMPERAGE CHARACTERISTICS

The arc-voltage and amperage characteristics of GMA welding are different from those for most other welding processes. The voltage is set on the welder, and the amperage is set by changing the wire-feed speed. At any one voltage setting the amperage required to melt the wire must change as it is fed into the weld. More amperage is required to melt the wire the faster it is fed, and less the slower it is fed.

Because changes in the wire-feed speed directly change the amperage, it is possible to set the amperage by using a chart and measuring the length of wire fed per minute, Table 2.1. The voltage and amperage required for a specific metal transfer method differ for various wire sizes, shielding gases, and metals.

The voltage and amperage setting will be specified for all welding done according to a welding procedure specification (WPS) or other codes and standards. However, most welding—like that done in small production shops, as maintenance welding, for repair work, in farm shops, and the like—is not done to a specific code or standard and therefore no specific setting exists. For that reason, it is important to learn to make the adjustments necessary to allow you to produce quality welds.

EXPERIMENT 2-2

Setting the Current

Module 1
Key Indicator 1, 2, 4

Module 5
Short-circuit Transfer
Key Indicator 4
Spray Transfer
Key Indicator 9

Using a properly assembled GMA welding machine, proper safety protection, and one piece of mild steel plate approximately 12 in. (305 mm) long × 1/4 in. (6 mm) thick, you will change the current settings and observe the effect on GMAW. On a scale of 0 to 10, set the wire-feed speed control dial at 5, or halfway between the low and high settings of

Table 2.1 Typical Amperages for Carbon Steel

Wire-feed Speed* (in./min)	Wire Diameter			
	.030 in. (0.8 mm)	.035 in. (0.9 mm)	.045 in. (1.2 mm)	.062 in. (1.6 mm)
100 (2.5)	40	65	120	190
200 (5.0)	80	120	200	330
300 (7.6)	130	170	260	425
400 (10.2)	160	210	320	490
500 (12.7)	180	245	365	–
600 (15.2)	200	265	400	–
700 (17.8)	215	280	430	–

*To check feed speed, run out wire for one minute and then measure its length.

the unit. The voltage is also set at a point halfway between the low and high settings. The shielding gas can be CO_2, argon, or a mixture. The gas flow should be adjusted to a rate of 35 cfh (16 L/min).

Hold the welding gun at a comfortable angle, lower your welding hood, and pull the trigger. As the wire feeds and contacts the plate, the weld will begin. Move the gun slowly along the plate. Note the following welding conditions as the weld progresses: voltage, amperage, weld direction, metal transfer, spatter, molten weld pool size, and penetration. Stop and record your observations in Table 2.2. Evaluate the quality of the weld as acceptable or unacceptable.

Reduce the voltage somewhat and make another weld, keeping all other weld variables (travel speed, stickout, direction, amperage) the same. Observe the weld and upon stopping record the results. Repeat this procedure until the voltage has been lowered to the minimum value indicated on the machine. Near the lower end the wire may stick, jump, or simply no longer weld.

Return the voltage indicator to the original starting position and make a short test weld. Stop and compare the results to those first observed. Then slightly increase the voltage setting and make another weld. Repeat the procedure of observing and recording the results as the voltage is increased in steps until the maximum machine capability is obtained. Near the maximum setting the spatter may become excessive if CO_2 shielding gas is used. Care must be taken to prevent the wire from fusing to the contact tube.

Return the voltage indicator to the original starting position and make a short test weld. Compare the results observed with those previously obtained.

Lower the wire-feed speed setting slightly and use the same procedure as before. First lower and then raise the voltage through a complete range and record your observations. After a complete set of test results are obtained from this amperage setting, again lower the wire-feed speed for a new series of tests. Repeat this procedure until the amperage is at the minimum setting shown on the machine. At low amperages and high voltage settings, the wire may tend to pop violently as a result of the uncontrolled arc.

Return the wire-feed speed and voltages to the original settings. Make a test weld and compare the results with the original tests. Slightly raise the wire speed and again run a set of tests as the voltage is changed in small steps. After each series, return the voltage setting to the starting point and increase the wire-feed speed. Make a new set of tests.

All of the test data can be gathered into an operational graph for the machine, wire type, size, and shielding gas. Set up a graph like that in Figure 2.19 to plot the data. The acceptable welds should be marked on

Table 2.2 Setting the Current

Weld Acceptability	Voltage	Amperage	Spatter	Molten Pool Size	Penetration
Good	20	75	Light	Small	Little

Electrode diameter	0.035 in. (0.9 mm)
Shielding gas CO_2	
Welding direction	Backhand

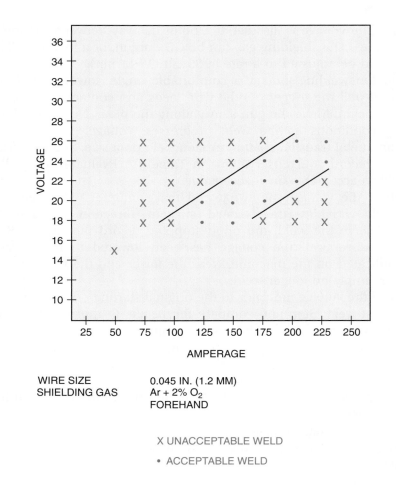

Figure 2.19 Graph for GMAW machine settings

the lines that extend from the appropriate voltages and amperages. Upon completion, the graph will give you the optimum settings for the operation of this particular GMAW setup. The optimum settings are along a line in the center of the acceptable welds.

Experienced welders will follow a much shorter version of this type of procedure anytime they are starting to work on a new machine or testing for a new job. This experiment can be repeated using different types of wire, wire sizes, shielding gases, and weld directions. Turn off the welding machine and shielding gas and clean up your work area when you are finished welding.

Complete a copy of the "Student Welding Report" listed in Appendix I or provided by your instructor.

ELECTRODE EXTENSION

Because of the constant-potential (CP) power supply, the welding current will change as the distance between the contact tube and the work changes. Although this change is slight, it is enough to affect the weld being produced. The longer the electrode extension, the greater the resistance will be to the welding current flowing through the small welding wire. This results in some of the welding current being changed to heat at the tip of the electrode, Figure 2.20. With a standard constant-current (CC) power supply for SMA welding, the heat buildup would also reduce the arc voltage, but with a CP power supply the voltage remains constant and the amperage increases. If the electrode extension is shortened, the welding current decreases.

The increase in current does not result in an increase in penetration, because the current is being used to heat the electrode tip and not being transferred to the weld metal. Penetration is reduced and buildup is increased as the electrode extension is lengthened. Penetration is increased and buildup decreased as the electrode extension is shortened. Controlling the weld penetration and buildup by changing the electrode will help maintain weld bead shape during welding. It will also help you better understand what may be happening if a weld starts out correctly but begins to change as it progresses along the joint. You may be changing the electrode extension without noticing the change. Short electrode stickout gives a hotter weld, and long stickout results in a cooler weld.

EXPERIMENT 2-3

Electrode Extension

Using a properly assembled GMA welding machine, proper safety protection, and a few pieces of mild steel, each about 12 in. (305 mm) long and ranging in thickness from 16 gauge to 1/2 in. (13 mm), you will observe the effect of changing electrode extension on the weld.

Start at a low current setting. Using the graph developed in Experiment 2.2, set both the voltage and amperage. The settings should be equal to those on the optimum line established for the wire type and size being used with the same shielding gas.

Holding the welding gun at a comfortable angle and height, lower your helmet and start to weld. Make a weld approximately 2 in. (51 mm) long. Then reduce the distance from the gun to the work while continuing to weld. After a few inches, again shorten the electrode extension even

Module 1
Key Indicator 1, 2, 4

Module 5
Short-circuit Transfer
Key Indicator 4
Spray Transfer
Key Indicator 9

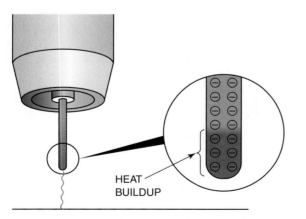

HEAT
BUILDUP

Figure 2.20 Heat buildup due to the extremely high current for the small conductor (electrode)

more. Keep doing this in steps until the nozzle is as close as possible to the work. Stop and return the gun to the original starting distance.

Repeat the process just described but now increase the electrode extension and make welds of a few inches at each extension. Keep increasing the electrode extension until the weld will no longer fuse or the wire becomes impossible to control.

Change the plate thickness and repeat the procedure. When the series has been completed with each plate thickness, raise the voltage and amperage to a medium setting and repeat the process. Upon completing this series of tests, adjust the voltage and amperage upward to a high setting. Make a full series of tests using the same procedures as before.

Record the results in Table 2.3 after each series of tests. The final results can be plotted on a graph, as was done in Figure 2.21, to establish the optimum electrode extension for each thickness, voltage, and amperage. Turn off the welding machine and shielding gas and clean up your work area when you are finished welding.

Complete a copy of the "Student Welding Report" listed in Appendix I or provided by your instructor.

Table 2.3 **Electrode Extension**

Weld Acceptability	Voltage	Amperage	Electrode Extension	Contact Tube to Work Distance	Bead Shape
Poor	20	100	1 in. (25 mm)	1-1/4 in. (31 mm)	Narrow, high, with little penetration

Electrode diameter: 0.035 in. (0.9 mm)
Shielding gas: CO_2
Welding direction: Forehand

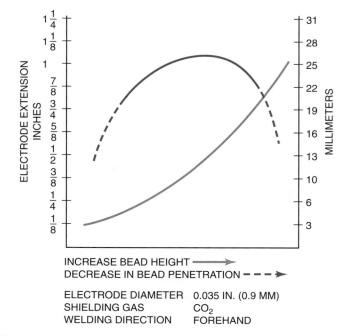

Figure 2.21 Plot of Experiment 2-3 results

WELDING GUN ANGLE

The term *welding gun angle* refers to the angle between the GMA welding gun and the work as it relates to the direction of travel. Backhand welding, or dragging angle, Figure 2.22, produces a weld with deep penetration and higher buildup. Forehand welding, or pushing angle, Figure 2.23, produces a weld with shallow penetration and little buildup.

Slight changes in the welding gun angle can be used to control the weld as the groove spacing changes. A narrow gap may require more penetration, but as the gap spacing increases a weld with less penetration may be required. Changing the electrode extension and welding gun angle at the same time can result in a quality weld being made under less than ideal conditions.

EXPERIMENT 2-4

Welding Gun Angle

Using a properly assembled GMA welding machine, proper safety protection, and some pieces of mild steel, each approximately 12 in. (305 mm) long and ranging in thickness from 16 gauge to 1/2 in. (13 mm), you will observe the effect of changing the welding gun angle on the weld bead.

Starting with a medium current setting and a plate that is 1/4 in. (6 mm) thick, hold the welding gun at a 30° angle to the plate in the direction of the weld, Figure 2.24. Lower your welding hood and depress the trigger. When the weld starts, move in a straight line and slowly pivot the gun angle as the weld progresses. Keep the travel speed, electrode

Module 1
Key Indicator 1, 2, 4

Module 5
Short-circuit Transfer
Key Indicator 4
Spray Transfer
Key Indicator 9

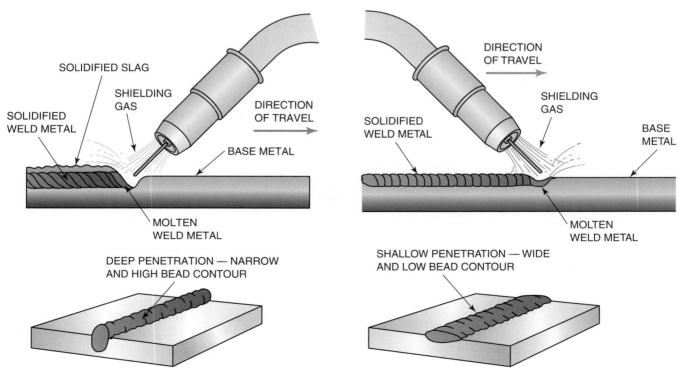

Figure 2.22 Backhand welding, or dragging angle

Figure 2.23 Forehand welding, or pushing angle

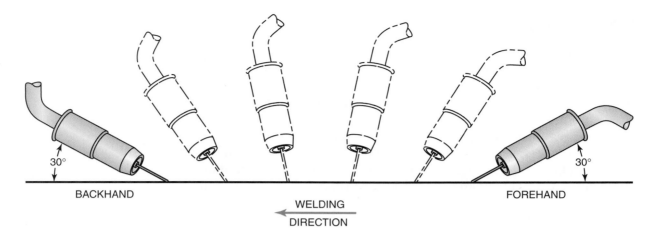

30°   BACKHAND

WELDING
DIRECTION

FOREHAND   30°

Figure 2.24 Welding gun angle

extension, and weave pattern (the side-to-side motion, if used) constant so that any change in the weld bead is caused by the angle change.

The pivot should be completed in the 12 in. (305 mm) of the weld. You will proceed from a 30° pushing angle to a 30° dragging angle. Repeat this procedure using different welding currents and plate thicknesses.

After the welds are complete, note the differences in width and reinforcement along the welds. Turn off the welding machine and shielding gas and clean up your work area when you are finished welding.

Complete a copy of the "Student Welding Report" listed in Appendix I or provided by your instructor.

EFFECT OF SHIELDING GAS ON WELDING

Shielding gases in the gas metal arc process are used primarily to protect the molten metal from oxidation and contamination. Other factors must be considered, however, in selecting the right gas for a particular application. Shielding gas can influence arc and metal transfer characteristics, weld penetration, width of fusion zone, surface shape patterns, welding speed, and undercut tendency. Inert gases such as argon and helium provide the necessary shielding because they do not form compounds with any other substance and are insoluble in molten metal. When used as pure gases for welding ferrous metals, argon and helium may produce an erratic arc action, promote undercutting, and result in other flaws.

It is therefore usually necessary to add controlled quantities of reactive gases to achieve good arc action and metal transfer with these materials. Adding oxygen or carbon dioxide to the inert gas tends to stabilize the arc, promote favorable metal transfer, and minimize spatter. As a result, the penetration pattern is improved and undercutting is reduced or eliminated.

Oxygen or carbon dioxide is often added to argon. The amount of reactive gas required to produce the desired effects is quite small. As little as 0.5% of oxygen will produce a noticeable change; 1% to 5% of oxygen is more common. Carbon dioxide may be added to argon in the 10% to 30% range. Mixtures of argon with less than 10% carbon dioxide may not have enough arc voltage to give the desired results.

Adding oxygen or carbon dioxide to an inert gas causes the shielding gas to become oxidizing. This in turn may cause porosity in some ferrous metals. In this case, a filler wire containing suitable deoxidizers should be used. The presence of oxygen in the shielding gas can also cause some loss of certain alloying elements, such as chromium, vanadium, aluminum, titanium, manganese, and silicon. Again, the addition of a deoxidizer to the filler wire is necessary.

Pure carbon dioxide has become widely used as a shielding gas for GMA welding of steels. It allows higher welding speed, better penetration in the short-circuiting transfer mode, and good mechanical properties, and it costs less than the inert gases. The chief drawback in the use of carbon dioxide is the less-steady-arc characteristics and considerable weld-metal-spatter losses. The spatter can be kept to a minimum by maintaining a very short, uniform arc length. Consistently sound welds can be produced using carbon dioxide shielding, provided that a filler wire having the proper deoxidizing additives is used.

EXPERIMENT 2-5

Effect of Shielding Gas Changes

Using a properly assembled GMA welding machine; proper safety protection; a source of CO_2, argon, and oxygen gases or a variety of premixed shielding gases; two flowmeters (or one two-gas mixing regulator); and some pieces of mild steel plate, each about 12 in. (305 mm) long and ranging in thickness from 16 gauge to 1/2 in. (13 mm), you will observe the effect of various shielding gas mixtures on the weld.

Using a mixing flowmeter regulator will allow the gases to be mixed in any desired mixture. A mixing ratio chart to be used to arrive at the approximate gas percentages appears in Figure 2.25. The exact ratios are not so important to you, as a student, as they are on code work.

Module 1
Key Indicator 1, 2, 4

Module 5
Short-circuit Transfer
Key Indicator 4
Spray Transfer
Key Indicator 9

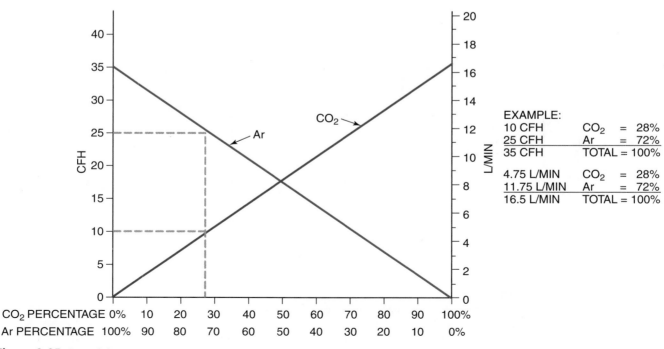

EXAMPLE:

10 CFH	CO_2	= 28%
25 CFH	Ar	= 72%
35 CFH	TOTAL	= 100%

4.75 L/MIN	CO_2	= 28%
11.75 L/MIN	Ar	= 72%
16.5 L/MIN	TOTAL	= 100%

Figure 2.25 Gas mixing percentages

With a medium voltage and amperage setting and using a 100% carbon dioxide (CO_2) shielding gas, start making a weld. Either change the mixture after each weld or have another person change the shielding gas during the weld. Keep the total flow rate the same by adding argon (Ar) while reducing the CO_2 to preserve the same flow rate. During the experiment, change over the shielding gas to 100% argon (Ar).

After the weld is complete, evaluate it for spatter, penetration, undercut, buildup, width, or other noticeable changes along its length. Using Table 2.4, record the results of your evaluation.

Repeat the procedure just explained two times more with both low and high power settings. Again, record your observations.

Starting with 100% argon (Ar), add oxygen (O_2) to the shielding gas. The oxygen percentage will range from 0% to 10%, Figure 2.26. Very slight changes in the percentage will have dramatic effects on the weld. You will make three welds using low, medium, and high power settings. For each weld, you will record your observations.

Table 2.4 Shielding Gas Mixtures

Weld Acceptability	Voltage	Spatter	Penetration	Puddle Size	Bead Appearance
Good	75 Ar 25 CO_2	Very little	Deep	Large	Wide with little buildup

Electrode diameter: 0.035 in. (0.9 mm)
Welding direction: Backhand
Voltage: 25
Amperage: 150

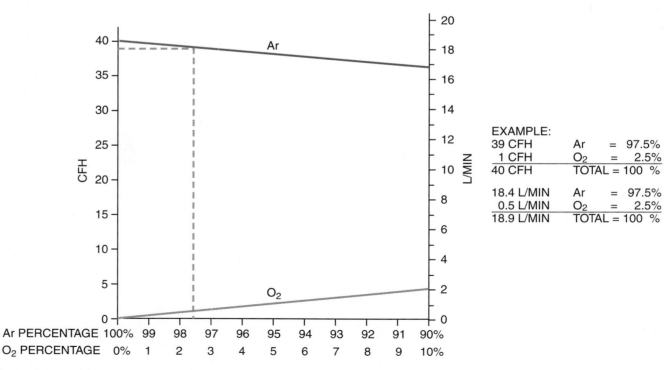

Figure 2.26 Ar and O_2 Mixture Percentages

During some of the welding tests, you will notice a change in the method of metal transfer, weld heat, and general weld performance without a change in the current settings. The shielding gas mixture can have major effects on the rate of metal transfer and the welding speed, as well as other welding variables. Higher speeds and greater production can be obtained by using some gas mixtures. However, the savings can be completely offset by the higher gas cost. Before making a final decision about the gas to be used, all the variables must be compared. Table 2.5 lists premixed shielding gases and their uses.

Turn off the welding machine and shielding gas and clean up your work area when you are finished welding.

Complete a copy of the "Student Welding Report" listed in Appendix I or provided by your instructor.

PRACTICES

The practices in this chapter are grouped according to those requiring similar techniques and setups. To make acceptable GMA welds consistently, the major skill required is the ability to set up the equipment and weldment. Changes such as variations in material thickness, position, and type of joint require changes in both technique and setup. A correctly set up GMA welding station can, in many cases, be operated with minimum skill. Often the only difference between a welder earning a minimum wage and one earning the maximum wage is the ability to make correct machine setups.

Ideally, only a few tests would be needed for the welder to make the necessary adjustments in setup and manipulation techniques to achieve

Table 2.5 Shielding Gases and Gas Mixtures Used for Gas Metal Arc Welding

Shielding Gas	Chemical Behavior	Uses and Usage Notes
1. Argon	Inert	Welding virtually all metals except steel
2. Helium	Inert	Al and Cu alloys for greater heat and to minimize porosity
3. Ar and He (20% to 80% to 50% to 50%)	Inert	Al and Cu alloys for greater heat and to minimize porosity but with quieter, more readily controlled arc action
4. N_2	Reducing	On Cu, very powerful arc
5. Ar + 25% to 30% N_2	Reducing	On Cu, powerful but smoother operating, more readily controlled arc than with N_2
6. Ar + 1% to 2% O_2	Oxidizing	Stainless and alloy steels, also for some deoxidized copper alloys
7. Ar + 3% to 5% O_2	Oxidizing	Plain carbon, alloy, and stainless steels (generally requires highly deoxidized wire)
8. Ar + 3% to 5% O_2	Oxidizing	Various steels using deoxidized wire
9. Ar + 20% to 30% O_2	Oxidizing	Various steels, chiefly with short-circuiting arc
10. Ar + 5% O_2 + 15% CO_2	Oxidizing	Various steels using deoxidized wire
11. CO_2	Oxidizing	Plain-carbon and low-alloy steels, deoxidized wire essential
12. CO_2 + 3% to 10% O_2	Oxidizing	Various steels using deoxidized wire
13. CO_2 + 20% O_2	Oxidizing	Steels

a good weld. The previous welding experiments should have given the welder a graphic set of comparisons to help that welder make the correct changes. In addition to keeping the test data, you may want to keep the test plates for a more accurate comparison.

The grouping of practices in this chapter will keep the number of variables in the setup to a minimum. Often, the only change required before going on to the next weld is to adjust the power settings.

Figures that are given in some of the practices will give the welder general operating conditions, such as voltage, amperage, and shielding gas or gas mixture. These are general values, so the welder will have to make some fine adjustments. Differences in the type of machine being used and the material surface condition will affect the settings. For this reason, it is preferable to use the settings developed during the experiments.

METAL PREPARATION

All hot-rolled steel has an oxide layer, which is formed during the rolling process, called mill scale. *Mill scale* is a thin layer of dark gray or black iron oxide. Some hot-rolled steels that have had this layer removed either mechanically or chemically can be purchased. However, almost all of the hot-rolled steel used today still has this layer because it offers some protection from rusting.

Mill scale is not removed for noncode welding, because it does not prevent most welds from being suitable for service. For practice welds that will be visually inspected, mill scale can usually be left on the plate. Filler metals and fluxes usually have deoxidizers added to them so that the adverse effects of the mill scale are reduced or eliminated, Table 2.6. But with GMA welding wire it is difficult to add enough deoxidizers to remove all effects of mill scale. The porosity that mill scale causes is most often confined to the interior of the weld and is not visible on the surface, Figure 2.27. Because it is not visible on the surface, it usually goes unnoticed and the weld passes visual inspection.

If the practice results are going to be destructively tested or if the work is of a critical nature, then all welding surfaces within the weld groove and the surrounding surfaces within 1 in. (25 mm) must be cleaned to bright metal, Figure 2.28. Cleaning may be either grinding, filing, sanding, or blasting.

Table 2.6 Deoxidizing Elements in Filler Wire

Deoxidizing Element	Strength
Aluminum (Al)	Very strong
Manganese (Mn)	Weak
Silicon (Si)	Weak
Titanium (Ti)	Very strong
Zirconium (Zr)	Very strong

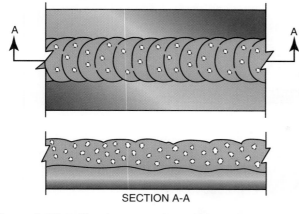

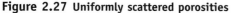

SECTION A-A

Figure 2.27 Uniformly scattered porosities

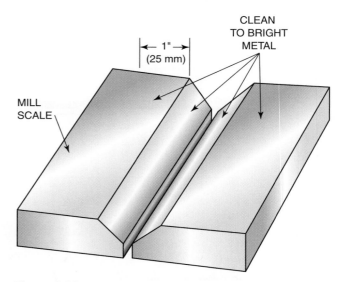

CLEAN
TO BRIGHT
METAL

1"
(25 mm)

MILL
SCALE

Figure 2.28 Clean all surfaces to bright metal before welding

FLAT POSITION, 1G AND 1F POSITIONS
PRACTICE 2-3

Stringer Beads Using the Short-circuiting Metal Transfer Method in the Flat Position

Using a properly set up and adjusted GMA welding machine, Table 2.7, proper safety protection, 0.035-in. and/or 0.045-in. (0.9-mm and/or 1.2-mm) diameter wire, and two or more pieces of mild steel sheet 12 in. (305 mm) long and 16 gauge and 1/8 in. (3 mm) thick, you will make a stringer bead weld in the flat position, Figure 2.29.

Starting at one end of the plate and using either a pushing or dragging technique, make a weld bead along the entire 12-in. (305-mm) length of the metal. After the weld is complete, check its appearance. Make any needed changes in voltage, wire feed speed, or electrode extension to correct the weld (refer to Figure 2.19 and Figure 2.21). Repeat the weld and make additional adjustments. After the machine is set, start to work on improving the straightness and uniformity of the weld.

Keeping the bead straight and uniform can be hard because of the limited visibility due to the small amount of light and the size of the molten weld pool. The welder's view is further restricted by the shielding gas nozzle, Figure 2.30. Even with limited visibility, it is possible to make a

Module 1
Key Indicator 1, 2, 3, 4

Module 2
Key Indicator 1, 2, 3, 4, 7

Module 5
Short-circuit Transfer
Key Indicator 4

Table 2.7 Typical Welding Current Settings for Short-circuiting Metal Transfer for Mild Steel

Process	Wire Diameter	Amperage Range (Optimum)	Voltage Range (Optimum)	Shielding Gas
Short-circuiting	0.030	60 (100) 140	14 (15) 16	100% CO_2
	0.035	90 (130) 150	16 (17) 20	75% Ar + 25% CO_2
				98% Ar + 2% O

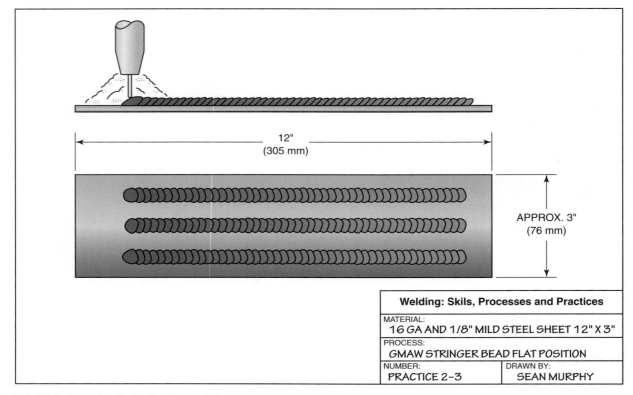

Figure 2.29 Stringer beads in the flat position

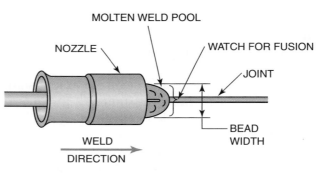

Figure 2.30 Watching the weld in forehand welding
The shielding gas nozzle restricts the welder's view of the weld bead when pushing.

satisfactory weld by watching the edge of the molten weld pool, the sparks, and the weld bead produced. Watching the leading edge of the molten weld pool (forehand welding, pushing technique) will show you the molten weld pool fusion and width. Watching the trailing edge of the molten weld pool (backhand welding, dragging technique) will show you the amount of buildup and the relative heat input, Figure 2.31. The quantity and size of sparks produced can indicate the relative location of the filler wire in the molten weld pool. The number of sparks will increase as the wire strikes the solid metal ahead of the molten weld pool. The gun itself will begin to vibrate or bump as the wire momentarily pushes against the cooler, unmelted base metal before it melts. Changes in weld width, buildup, and proper joint tracking can be seen by watching the bead as it appears from behind the shielding gas nozzle.

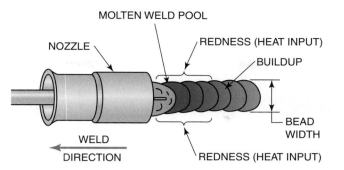

MOLTEN WELD POOL

NOZZLE

REDNESS (HEAT INPUT)

BUILDUP

BEAD WIDTH

WELD DIRECTION

REDNESS (HEAT INPUT)

Figure 2.31 Watching the weld in backhand welding
Watch the trailing edge of the molten weld pool.

Repeat each type of bead as needed until consistently good beads are obtained. Turn off the welding machine and shielding gas and clean up your work area when you are finished welding.

Complete a copy of the "Student Welding Report" listed in Appendix I or provided by your instructor.

PRACTICE 2-4

Flat Position Butt Joint, Lap Joint, and Tee Joint

Using the same equipment, materials, and procedures listed in Practice 2-3, make welded butt joints, lap joints, and tee joints in the flat position, Figure 2.32A, 2.32B and 2.32C.

- Tack weld the sheets together and place them flat on the welding table, Figure 2.33.

Module 1
Key Indicator 1, 2, 3, 4

Module 2
Key Indicator 1, 2, 3, 4, 7

Module 5
Key Indicator 5, 6
This exercise addresses the "Flat" position portion of the all-position requirement.

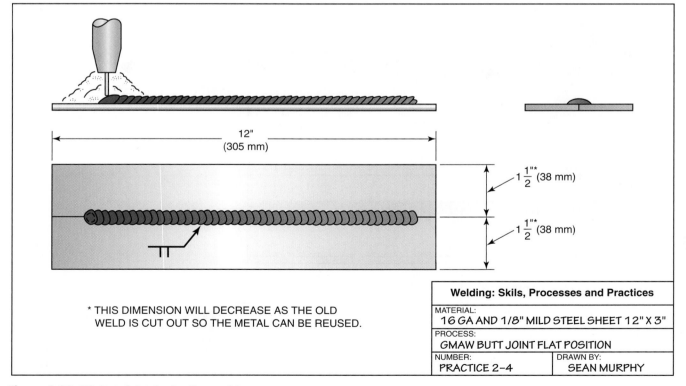

12" (305 mm)

$1\frac{1}{2}$"* (38 mm)

$1\frac{1}{2}$"* (38 mm)

* THIS DIMENSION WILL DECREASE AS THE OLD WELD IS CUT OUT SO THE METAL CAN BE REUSED.

Welding: Skils, Processes and Practices

MATERIAL: 16 GA AND 1/8" MILD STEEL SHEET 12" X 3"

PROCESS: GMAW BUTT JOINT FLAT POSITION

NUMBER: PRACTICE 2-4   DRAWN BY: SEAN MURPHY

Figure 2.32 (A) Butt joint in the flat position

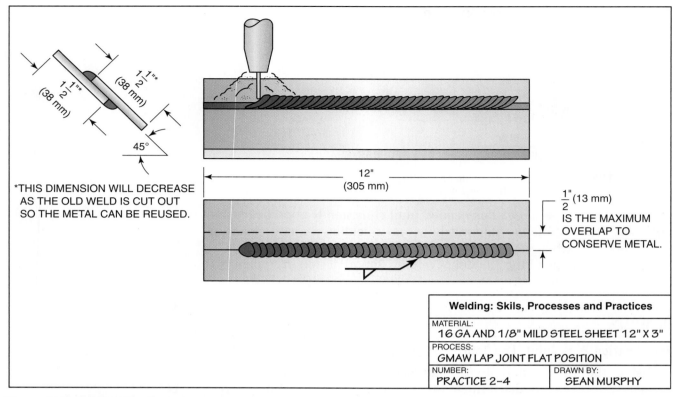

*THIS DIMENSION WILL DECREASE AS THE OLD WELD IS CUT OUT SO THE METAL CAN BE REUSED.

$\frac{1}{2}$" (13 mm) IS THE MAXIMUM OVERLAP TO CONSERVE METAL.

Welding: Skils, Processes and Practices	
MATERIAL: 16 GA AND 1/8" MILD STEEL SHEET 12" X 3"	
PROCESS: GMAW LAP JOINT FLAT POSITION	
NUMBER: PRACTICE 2–4	DRAWN BY: SEAN MURPHY

Figure 2.32 (B) Lap joint in the flat position

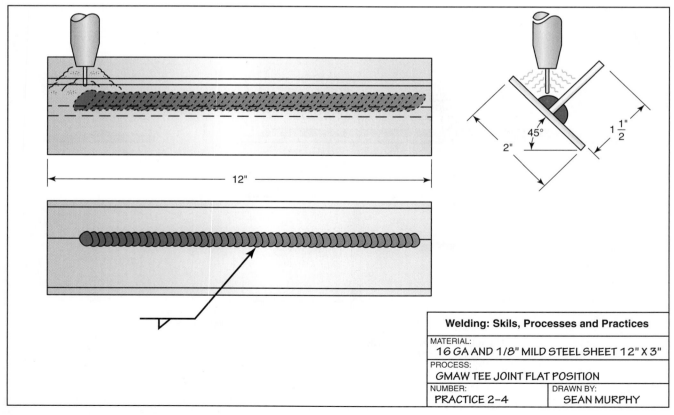

Welding: Skils, Processes and Practices	
MATERIAL: 16 GA AND 1/8" MILD STEEL SHEET 12" X 3"	
PROCESS: GMAW TEE JOINT FLAT POSITION	
NUMBER: PRACTICE 2–4	DRAWN BY: SEAN MURPHY

Figure 2.32 (C) Tee joint in the flat position

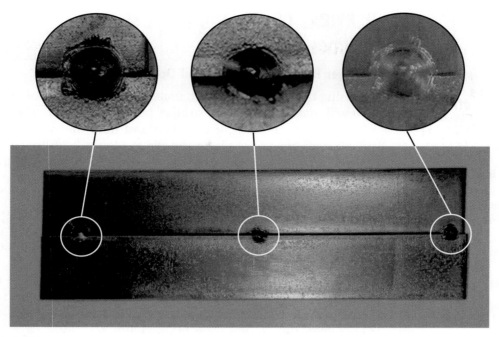

Figure 2.33 Tack welds
Use enough tack welds to keep the joint in alignment during welding. Small tack welds are easier to weld over without adversely affecting the weld.
Source: Courtesy of Larry Jeffus

- Starting at one end, run a bead along the joint. Watch the molten weld pool and bead for signs that a change in technique may be required.
- Make any needed changes as the weld progresses. By the time the weld is complete, you should be making the weld nearly perfectly.
- Using the same technique that was established in the last weld, make another weld. This time, the entire 12 in. (305 mm) of weld should be flawless.

Repeat each type of joint with both thicknesses of metal until consistently good beads are obtained. Turn off the welding machine and shielding gas and clean up your work area when you are finished welding.

Complete a copy of the "Student Welding Report" listed in Appendix I or provided by your instructor.

PRACTICE 2-5

Flat Position Butt Joint, Lap Joint, and Tee Joint, All with 100% Penetration

Using the same equipment, materials, and setup listed in Practice 2-3, make a welded joint in the flat position with 100% penetration along the entire 12-in. (305-mm) length of the welded joint. Repeat each type of joint until consistently good beads are obtained. Turn off the welding machine and shielding gas and clean up your work area when you are finished welding.

Complete a copy of the "Student Welding Report" listed in Appendix I or provided by your instructor.

Module 1
Key Indicator 1, 2, 3, 4

Module 2
Key Indicator 1, 2, 3, 4, 7

Module 5
Key Indicator 5, 6
This exercise addresses the "Flat" position portion of the all-position requirement.

VERTICAL UP 3G AND 3F POSITIONS
PRACTICE 2-6

Stringer Bead at a 45° Vertical Up Angle

Using the same equipment, materials, and setup as listed in Practice 2-3, you will make a vertical up stringer bead on a plate at a 45° inclined angle.

Start at the bottom of the plate and hold the welding gun at a slight pushing or upward angle to the plate, Figure 2.34. Brace yourself, lower your hood, and begin to weld. Depending upon the machine settings and type of shielding gas used, you will make a weave pattern.

If the molten weld pool is large and fluid (hot), use a "C" or "J" weave pattern to allow a longer time for the molten weld pool to cool, Figure 2.35. Do not make the weave so long or fast that the wire is allowed to strike the metal ahead of the molten weld pool. If this happens, spatter increases and a spot or zone of incomplete fusion may occur, Figure 2.36.

If the molten weld pool is small and controllable, use a small "C," zig-zag, or "J" weave pattern to control the width and buildup of the weld. A slower speed can also be used. Watch for complete fusion along the leading edge of the molten weld pool. Figure 2.37 shows a weld that did not fuse with the plate.

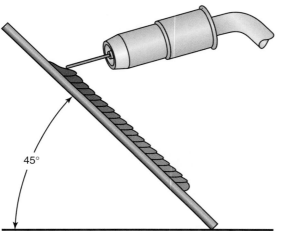

45°

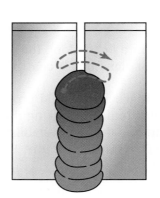
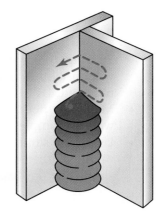

Figure 2.35 Vertical up weld patterns
Left, "C" pattern; right, "J" pattern.

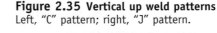
SHELF

Figure 2.34 Vertical up position
Source: Photo courtesy of Larry Jeffus

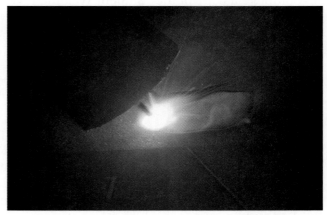

Figure 2.36 Burst of spatter caused by incorrect electrode contact with base metal
Source: Courtesy of Larry Jeffus

Figure 2.37 Weld separated from the plate
There is no fusion between the weld and plate.
Source: Courtesy of Larry Jeffus

A weld that is high and has little or no fusion is too "cold." Changing the welding technique will not correct this problem. The welder must stop welding and make the needed adjustments.

As the weld progresses up the plate, the back or trailing edge of the molten weld pool will cool, forming a shelf to support the molten metal. Watch the shelf to be sure that molten metal does not run over, forming a drip. When it appears that the metal may flow over the shelf, either increase the weave lengths, lengthen the electrode extension, or stop and start the current for brief moments to allow the weld to cool. Stopping for brief moments will not allow the shielding gas to be lost.

Continue to weld along the entire 12-in. (305-mm) length of the plate. Repeat this weld until a straight and uniform weld bead is produced. Turn off the welding machine and shielding gas and clean up your work area when you are finished welding.

Complete a copy of the "Student Welding Report" listed in Appendix I or provided by your instructor.

PRACTICE 2-7

Stringer Bead in the Vertical Up Position

Repeat Practice 2-6 and increase the angle of the plate until you have mastered a straight and uniform weld bead in the vertical up position. Turn off the welding machine and shielding gas and clean up your work area when you are finished welding.

Complete a copy of the "Student Welding Report" listed in Appendix I or provided by your instructor.

PRACTICE 2-8

Butt Joint, Lap Joint, and Tee Joint in the Vertical Up Position at a 45° Angle

Using the same equipment, materials, and setup as listed in Practice 2-3, you will make vertical up welded joints on a plate at a 45° inclined angle.

Tack weld the metal pieces together and brace them in position. Check to see that you have free movement along the entire joint to

Module 1
Key Indicator 1, 2, 3, 4

Module 2
Key Indicator 1, 2, 3, 4, 7

Module 5
Short-circuit Transfer
Key Indicator 4

Module 1
Key Indicator 1, 2, 3, 4

Module 2
Key Indicator 1, 2, 3, 4, 7

Module 5
Key Indicator 5, 6
This exercise addresses the "Vertical" position portion of the all-position requirement.

prevent stopping and restarting during the weld. Avoiding stops and starts both speeds up the welding time and eliminates discontinuities.

The weave pattern should allow for adequate fusion on both edges of the joint. Watch the edges to be sure that they are being melted so that adequate fusion and penetration occur.

Repeat each type of joint as needed until consistently good beads are obtained. Turn off the welding machine and shielding gas and clean up your work area when you are finished welding.

Complete a copy of the "Student Welding Report" listed in Appendix I or provided by your instructor.

PRACTICE 2-9

Module 1
Key Indicator 1, 2, 3, 4

Module 2
Key Indicator 1, 2, 3, 4, 7

Module 5
Key Indicator 5, 6
This exercise addresses the "Vertical" position portion of the all-position requirement.

Butt Joint, Lap Joint, and Tee Joint in the Vertical Up Position with 100% Penetration

Using the same equipment, materials, and setup as listed in Practice 2-3, you will increase the plate angle gradually as you develop skill until you are making satisfactory welds in the vertical up position.

Repeat each type of joint as needed until consistently good beads are obtained. Turn off the welding machine and shielding gas and clean up your work area when you are finished welding.

Complete a copy of the "Student Welding Report" listed in Appendix I or provided by your instructor.

VERTICAL DOWN 3G AND 3F POSITIONS

The vertical down welding technique can be useful when making some types of welds. The major advantages of this technique are the following:

- Speed—Very high rates of travel are possible.
- Shallow penetration—Thin sections or root openings can be welded with little burn-through.
- Good bead appearance—The weld has a nice width-to-height ratio and is uniform.

Vertical down welds are often used on thin sheet metals or in the root pass in grooved joints. The combination of controlled penetration and higher welding speeds makes vertical down the best choice for such welds. The ease with which welds having a good appearance can be made is deceiving. Generally, more skill is required to make sound welds with this technique than in the vertical up position. The most common problem with these welds is lack of fusion or overlap. To prevent these problems, the arc must be kept at or near the leading edge of the molten weld pool.

Module 1
Key Indicator 1, 2, 3, 4

Module 2
Key Indicator 1, 2, 3, 4, 7

Module 5
Short-Circuit Transfer
Key Indicator 4

PRACTICE 2-10

Stringer Bead at a 45° Vertical Down Angle

Using the same equipment, materials, and setup as listed in Practice 2-3, you will make a vertical down stringer bead on a plate at a 45° inclined angle.

Holding the welding gun at the top of the plate with a slight dragging angle, Figure 2.38, will help to increase penetration, hold back the molten weld pool, and improve visibility of the weld. Be sure that your movements along the 12-in. (305-mm) length of plate are unrestricted.

Lower your hood and start the weld. Watch both the leading edge and the sides of the molten weld pool for fusion. The leading edge should flow into the base metal, not curl over it. The sides of the molten weld pool should also show fusion into the base metal and not be flashed (ragged) along the edges.

The weld may be made with or without a weave pattern. If a weave pattern is used, it should be a "C" pattern. The "C" should follow the leading edge of the weld. Some changes on the gun angle may help to increase penetration. Experiment with the gun angle as the weld progresses.

Repeat these welds until you have established a rhythm and technique that work well for you. The welds must be straight and uniform and have complete fusion. Turn off the welding machine and shielding gas and clean up your work area when you are finished welding.

Complete a copy of the "Student Welding Report" listed in Appendix I or provided by your instructor.

Figure 2.38 Vertical down position
Source: Courtesy of Larry Jeffus

PRACTICE 2-11

Stringer Bead in the Vertical Down Position

Repeat Practice 2-10 and increase the angle of the plate until you have developed the skill to repeatedly make good welds in the vertical down position. The weld bead must be straight and uniform and have complete fusion. Turn off the welding machine and shielding gas and clean up your work area when you are finished welding.

Complete a copy of the "Student Welding Report" listed in Appendix I or provided by your instructor.

PRACTICE 2-12

Butt Joint, Lap Joint, and Tee Joint in the Vertical Down Position

Using a properly set up and adjusted GMA welding machine, Table 2.7, proper safety protection, 0.035-in. and/or 0.045-in. (0.9-mm and/or 1.2-mm) diameter wire, and two or more pieces of mild steel sheet 12 in. (305 mm) long and 16 gauge and 1/8 in. (3 mm) thick, you will make vertical down welded joints.

Holding the welding gun at the top of the plate with a slight dragging angle, Figure 2.38, will help to increase penetration, hold back the molten weld pool, and improve visibility of the weld. Be sure that your movements along the 12-in. (305-mm) length of plate are unrestricted.

Lower your hood and start the weld. Watch both the leading edge and sides of the molten weld pool for fusion. The leading edge should flow into the base metal, not curl over it. The sides of the molten weld pool should also show fusion into the base metal and not be flashed (ragged) along the edges.

The weld may be made with or without a weave pattern. If a weave pattern is used, it should be a "C" pattern. The "C" should follow the leading edge of the weld. Some changes on the gun angle may help to increase penetration. Experiment with the gun angle as the weld progresses.

Repeat these welds until you have established a rhythm and technique that work well for you. The welds must be straight and uniform and have complete fusion. Turn off the welding machine and shielding gas and clean up your work area when you are finished welding.

Complete a copy of the "Student Welding Report" listed in Appendix I or provided by your instructor.

PRACTICE 2-13

Butt Joint and Tee Joint in the Vertical Down Position with 100% Penetration

Using the same equipment, materials, and setup as listed in Practice 2-3, you will make welded joints with 100% weld penetration.

It may be necessary to adjust the root opening to meet the penetration requirements. The lap joint was omitted from this practice because little additional skill can be developed with it that is not already acquired with the tee joint. Repeat each type of joint until consistently good welds are obtained. Turn off the welding machine and shielding gas and clean up your work area when you are finished welding.

Complete a copy of the "Student Welding Report" listed in Appendix I or provided by your instructor.

HORIZONTAL 2G AND 2F POSITIONS
PRACTICE 2-14

Horizontal Stringer Bead at a 45° Angle

Using the same equipment, materials, and setup as listed in Practice 2-3, you will make a horizontal stringer bead on a plate at a 45° reclined angle.

Start at one end with the gun pointed in a slightly upward direction, Figure 2.39. You may use a pushing or a dragging (a leading or a trailing) gun angle, depending upon the current setting and penetration desired. Undercutting along the top edge and overlap along the bottom edge are problems with both gun angles. Careful attention must be paid to the manipulation "weave" technique used to overcome these problems.

The most successful weave patterns are the "C" and "J" patterns. The "J" pattern is the most frequently used. It allows weld metal to be deposited along a shelf created by the previous weave, Figure 2.40. The length of the "J" can be changed to control the weld bead size. Smaller weld beads are easier to control than large ones.

Repeat these welds until you have established the rhythm and technique that work well for you. The weld must be straight and uniform and have complete fusion. Turn off the welding machine and shielding gas and clean up your work area when you are finished welding.

Complete a copy of the "Student Welding Report" listed in Appendix I or provided by your instructor.

Module 1
Key Indicator 1, 2, 3, 4

Module 2
Key Indicator 1, 2, 3, 4, 7

Module 5
Short-circuit Transfer
Key Indicator 4

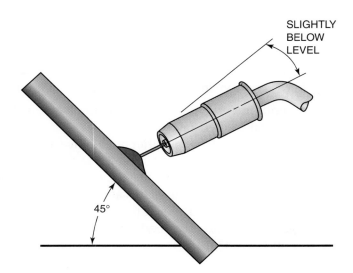

SLIGHTLY
BELOW
LEVEL

45°

Figure 2.39 45° horizontal position

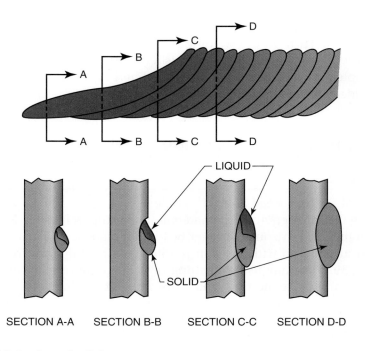

Figure 2.40 Sections of a "J" weave pattern
The actual size of the molten weld pool remains small along the weld.

PRACTICE 2-15

Stringer Bead in the Horizontal Position

Repeat Practice 2-14 and increase the angle of the plate until you have developed the skill to repeatedly make good horizontal welds on a vertical surface. The weld bead must be straight and uniform and have complete fusion. Turn off the welding machine and shielding gas and clean up your work area when you are finished welding.

Complete a copy of the "Student Welding Report" listed in Appendix I or provided by your instructor.

PRACTICE 2-16

Butt Joint, Lap Joint, and Tee Joint in the Horizontal Position

Using the same equipment, materials, and setup listed in Practice 2-3, you will make horizontal welded joints.

Tack weld the pieces of metal together and brace them in position using the same skills developed in Practice 2-14. Starting at one end, make a weld along the entire length of the joint. When making the butt or lap joints, it may help to recline the plates at a 45° angle until you have developed the technique required. Repeat each type of joint as needed until consistently good welds are obtained. Turn off the welding machine and shielding gas and clean up your work area when you are finished welding.

Complete a copy of the "Student Welding Report" listed in Appendix I or provided by your instructor.

PRACTICE 2-17

Butt Joint and Tee Joint in the Horizontal Position with 100% Penetration

Using the same equipment, materials, and setup as listed in Practice 2-3, you will make overhead joints having 100% penetration in the horizontal position.

It may be necessary to adjust the root opening to meet the penetration requirements. Repeat each type of joint as needed until consistently good welds are obtained. Turn off the welding machine and shielding gas and clean up your work area when you are finished welding.

Complete a copy of the "Student Welding Report" listed in Appendix I or provided by your instructor.

Module 1
Key Indicator 1, 2, 3, 4

Module 2
Key Indicator 1, 2, 3, 4, 7

Module 5
Key Indicator 5, 6
This exercise addresses the "Horizontal" position portion of the all-position requirement.

OVERHEAD 4G AND 4F POSITIONS

There are several advantages to the use of short-circuiting arc metal transfer in the overhead position, including the following:

- Small molten weld pool size—The smaller size of the molten weld pool allows surface tension to hold it in place. Less molten weld pool sag results in improved bead contour with less undercut and fewer icicles, Figure 2.41.
- Direct metal transfer—The direct metal transfer method does not rely on other forces to get the filler metal into the molten weld pool. This results in efficient metal transfer and less spatter and loss of filler metal.

PRACTICE 2-18

Stringer Bead in the Overhead Position

Using the same equipment, materials, and setup as listed in Practice 2-3, you will make a welded stringer bead in the overhead position.

The molten weld pool should be kept as small as possible for easier control. A small molten weld pool can be achieved by using lower current settings, by using a longer wire stickout, by traveling faster, or by pushing the molten weld pool. The technique used is the welder's choice. Often a combination of techniques can be used with excellent results.

Module 1
Key Indicator 1, 2, 3, 4

Module 2
Key Indicator 1, 2, 3, 4, 7

Module 5
Short-circuit Transfer
Key Indicator 4

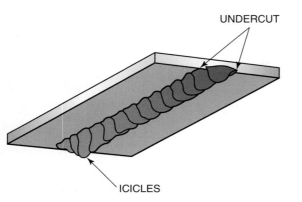

Figure 2.41 Overhead weld

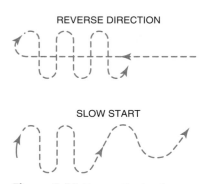

REVERSE DIRECTION

SLOW START

Figure 2.42 Two methods of concentrating heat at the beginning of a weld bead to aid in penetration depth

Lower current settings require closer control of gun manipulation to ensure that the wire is fed into the molten weld pool just behind the leading edge. The low power will cause overlap and more spatter if this contact position of wire to molten weld pool is not closely maintained.

Faster travel speeds allow the welder to maintain a high production rate even if multiple passes are required to complete the weld. Weld penetration into the base metal at the start of the bead can be obtained by using a slow start or quickly reversing the weld direction. Both the slow start and reversal of weld direction put more heat into the start to increase penetration, Figure 2.42. The higher speed also reduces the amount of weld distortion by reducing the amount of time that heat is applied to a joint.

The pushing, or trailing, gun angle forces the bead to be flatter by spreading it out over a wider area as compared to the bead resulting from a dragging, or backhand, gun angle. The wider, shallow molten weld pool cools faster, resulting in less time for sagging and the formation of icicles.

When welding overhead, extra personal protection is required to reduce the danger of burns. Wear leather sleeves or a leather jacket, and a cap.

Much of the spatter created during overhead welding falls into the shielding gas nozzle. The effectiveness of the shielding gas is reduced, Figure 2.43, and the contact tube may short out to the gas nozzle, Figure 2.44. Turbulence caused by the spatter obstructing the gas may lead to weld contamination. The shorted gas nozzle may arc to the work, causing damage both to the nozzle and to the plate. To control the amount of spatter, a longer stickout and/or a sharper gun-to-plate angle is required to allow most of the spatter to fall clear of the gas nozzle. The nozzle can be dipped, sprayed, or injected automatically, Figure 2.45, with antispatter to help prevent the spatter from sticking. Applying antispatter will not stop the spatter from building up, but it does make its removal much easier. Ensure that the antispatter gel does not restrict the flow of shield gas from the nozzle.

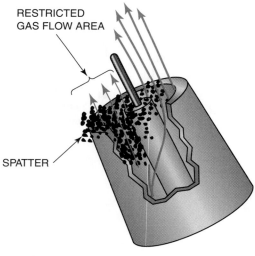

RESTRICTED GAS FLOW AREA

SPATTER

Figure 2.43 Shielding gas flow affected by excessive weld spatter in nozzle

Figure 2.44 Gas nozzle damaged after shorting out against the work
Source: Courtesy of Larry Jeffus

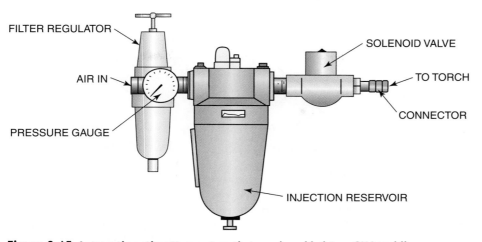

ANTISPATTER UNIT - MODEL 4050

FILTER REGULATOR

SOLENOID VALVE

AIR IN

TO TORCH

PRESSURE GAUGE

CONNECTOR

INJECTION RESERVOIR

Figure 2.45 Automatic antispatter system that can be added to a GMA welding gun

Make several short weld beads using various techniques to establish the method that is most successful and most comfortable for you. After each weld, stop and evaluate it before making a change. When you have decided on the technique to be used, make a welded stringer bead that is 12 in. (305 mm) long.

Repeat the weld until it can be made straight, uniform, and free from any visual defects. Turn off the welding machine and shielding gas and clean up your work area when you are finished welding.

Complete a copy of the "Student Welding Report" listed in Appendix I or provided by your instructor.

PRACTICE 2-19

Butt Joint, Lap Joint, and Tee Joint in the Overhead Position

Using the same equipment, materials, and setup as listed in Practice 2-3, you will make an overhead welded joint.

Tack weld the pieces of metal together and secure them in the overhead position. Be sure you have an unrestricted view and freedom of movement along the joint. Start at one end and make a weld along the joint. Use the same technique developed in Practice 2-18.

Repeat the weld until it can be made straight, uniform, and free from any visual defects. Turn off the welding machine and shielding gas and clean up your work area when you are finished welding.

Complete a copy of the "Student Welding Report" listed in Appendix I or provided by your instructor.

Module 1
Key Indicator 1, 2, 3, 4

Module 2
Key Indicator 1, 2, 3, 4, 7

Module 5
Key Indicator 5, 6
This exercise addresses the "Overhead" position portion of the all-position requirement.

PRACTICE 2-20

Butt Joint and Tee Joint in the Overhead Position with 100% Penetration

Using the same equipment, materials, and setup as listed in Practice 2-3, you will make overhead welded joints having 100% penetration.

Tack weld the metal together. It may be necessary to adjust the root opening to allow 100% weld metal penetration. During these welds, it may be necessary to use a dragging, or backhand, torch angle. When

Module 1
Key Indicator 1, 2, 3, 4

Module 2
Key Indicator 1, 2, 3, 4, 7

Module 5
Key Indicator 5, 6
This exercise addresses the "Overhead" position portion of the all-position requirement.

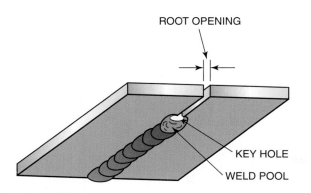

Figure 2.46 Overhead welding

used with a "C" or "J" weave pattern, this torch angle helps to achieve the desired depth of penetration. A key hole just ahead of the molten weld pool is a good sign that the metal is being penetrated, Figure 2.46.

Repeat the weld until it can be made straight, uniform, and free from any visual defects. Turn off the welding machine and shielding gas and clean up your work area when you are finished welding.

Complete a copy of the "Student Welding Report" listed in Appendix I or provided by your instructor.

AXIAL SPRAY TRANSFER
PRACTICE 2-21

Module 1
Key Indicator 1, 2, 3, 4

Module 2
Key Indicator 1, 2, 3, 4, 7

Module 5
Spray Transfer
Key Indicator 9

Stringer Bead, 1G Position

Using a properly set up and adjusted GMA welding machine (see Table 2.8), proper safety protection, 0.035-in. and/or 0.045-in. (0.9-mm and/or 1.2-mm)-diameter wire, and two or more pieces of mild steel plate 12 in. (305 mm) long × 1/4 in. (6 mm) thick, you will make a welded stringer bead in the flat position.

Start at one end of the plate and use either a push or drag technique to make a weld bead along the entire 12-in. (305-mm) length of the metal using spray or pulsed-arc metal transfer. After the weld is complete, check its appearance and make any changes needed to correct the weld, Figure 2.47. Repeat the weld and make any additional adjustments required. After the machine is set, start working on improving the straightness and uniformity of the weld. Turn off the welding machine and shielding gas and clean up your work area when you are finished welding.

Complete a copy of the "Student Welding Report" listed in Appendix I or provided by your instructor.

Table 2.8 Typical Welding Current Settings for Axial Spray Metal Transfer for Mild Steel

Process	Wire Diameter	Amperage Range (Optimum)	Voltage Range (Optimum)	Shielding Gas
Axial spray	0.030	115–200	15–27	98% Ar + 2% O_2
	0.035	165–300	18–32	
	0.045	200–450	20–34	

Figure 2.47 Weld bead made with GMAW axial spray metal transfer
Source: Courtesy of Larry Jeffus

PRACTICE 2-22

Butt Joint, Lap Joint, and Tee Joint Using the Spray Transfer Method

Using the same equipment, materials, and setup as listed in Practice 2-21, you will make a flat and horizontal weld using spray transfer or pulsed-spray metal transfer, Figure 2.48.

Tack weld the metal together and place the assembly in the flat position on the welding table. Start at one end and make a uniform weld along the entire 12-in. (305-mm) length of the joint. Watch the sides of the fillet weld for signs of undercutting.

Repeat the weld until it can be made straight, uniform, and free from any visual defects. Turn off the welding machine and shielding gas and clean up your work area when you are finished welding.

Complete a copy of the "Student Welding Report" listed in Appendix I or provided by your instructor.

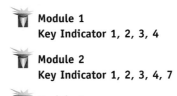

Module 1
Key Indicator 1, 2, 3, 4

Module 2
Key Indicator 1, 2, 3, 4, 7

Module 5
Spray Transfer
Key Indicator 10, 11

PRACTICE 2-23

Butt Joint and Tee Joint

Using the same equipment, materials, and setup as listed in Practice 2-21, you will make a flat weld using spray transfer. Each weld must pass the

Module 1
Key Indicator 1, 2, 3, 4

Module 2
Key Indicator 1, 2, 3, 4, 7

Module 5
Spray Transfer
Key Indicator 10, 11

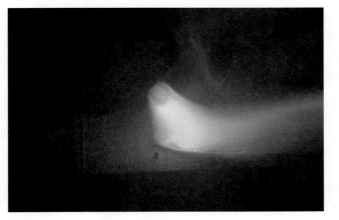

Figure 2.48 GMAW axial spray metal transfer
Source: Courtesy of Larry Jeffus

guided bend test. Repeat each type of weld joint as needed until the bend test can be passed. Turn off the welding machine and shielding gas and clean up your work area when you are finished welding.

Complete a copy of the "Student Welding Report" listed in Appendix I or provided by your instructor.

PRACTICE 2-24

Gas Metal Arc Welding—Short-Circuit Metal Transfer (GMAW-S) Workmanship Sample

Welding Procedure Specification (WPS)

Welding Procedure Specification No.: Practice 2-24 Date:

Title:

Welding GMAW-S of plate to plate.

Scope:

This procedure is applicable for square groove and fillet welds within the range of 10 ga. (3.4 mm) through 14 ga. (1.9 mm).

Welding may be performed in the following positions: all.

Base Metal:

The base metal shall conform to carbon steel M-1, P-1, and S-1 Group 1 or 2.

Backing material specification: none.

Filler Metal:

The filler metal shall conform to AWS specification no. E70S-X from AWS specification A5.18. This filler metal falls into F-number F-6 and A-number A-1.

Shielding Gas:

The shielding gas, or gases, shall conform to the following compositions and purity:

CO_2 at 30 to 50 cfh or 75% Ar/25% CO_2 at 30 to 50 cfh.

Joint Design and Tolerances:

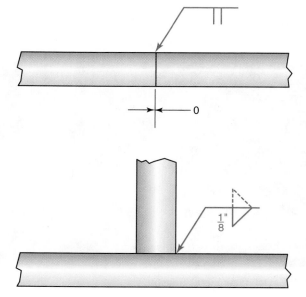

Module 1
Key Indicator 1, 2, 3, 4

Module 2
Key Indicator 1, 2, 3, 4, 7

Module 3
Key Indicator 1, 2, 3

Module 5
Short-circuit Transfer
Key Indicator 3, 4, 5, 6, 7

Module 9
Key Indicator 1, 2

Preparation of Base Metal:

All parts may be mechanically cut or machine PAC unless specified as manual PAC.

All hydrocarbons and other contaminations, such as cutting fluids, grease, oil, and primers, must be cleaned off all parts and filler metals before welding. This cleaning can be done with any suitable solvents or detergents. The groove face and inside and outside plate surface within 1 in. (25 mm) of the joint must be mechanically cleaned of slag, rust, and mill scale. Cleaning must be done with a wire brush or grinder down to bright metal.

Electrical Characteristics:

The current shall be direct-current electrode positive (DCEP). The base metal shall be on the negative side of the line.

Electrode		Welding Power			Shielding Gas		Base Metal	
Type	Size	Amps	Wire-feed Speed, ipm (cm/min)	Volts	Type	Flow	Type	Thickness
E70S-X	0.035 in. (0.9 mm)	90 to 120	180 to 300 (457 to 762)	15 to 19	CO_2 or 75% Ar/ CO_2 25%	30 to 50	Low-carbon steel	1/4 in. to 1/2 in. (6 mm to 13 mm)
E70S-X	0.045 in. (1.2 mm)	130 to 200	125 to 200 (318 to 508)	17 to 20	CO_2 or 75% Ar/ CO_2 25%	30 to 50	Low-carbon steel	1/4 in. to 1/2 in. (6 mm to 13 mm)

Preheat:

The parts must be heated to a temperature higher than 50°F (10°C) before any welding is started.

Backing Gas:

N/A

Safety:

Proper protective clothing and equipment must be used. The area must be free of all hazards that may affect the welder or others in the area. The welding machine, welding leads, work clamp, electrode holder, and other equipment must be in safe working order.

Welding Technique:

Using a 1/2-in. (13-mm) or larger gas nozzle for all welding, first tack weld the plates together according to the drawing. Use the E70S-X filler metal to fuse the plates together. Clean any silicon slag, being sure to remove any trapped silicon slag along the sides of the weld.

Using the E70S-X arc welding electrodes, make a series of stringer beads, no thicker than 3/16 in. (4.7 mm). The 1/8-in. (3.1-mm) fillet welds are to be made with one pass. All welds must be placed in the orientation shown in the drawing.

Interpass Temperature:

The plate should not be heated to a temperature higher than 350°F (175°C) during the welding process. After each weld pass is completed, allow it to cool but never to a temperature below 50°F (10°C). The weldment must not be quenched in water.

Cleaning:

Any slag must be cleaned off between passes. The weld beads may be cleaned by a hand wire brush, a hand chipping, a punch and hammer, or a needle-scaler. All weld cleaning must be performed with the test plate in the welding position.

Visual Inspection:

Visually inspect the weld for uniformity and discontinuities.

1. There shall be no cracks, no incomplete fusion.
2. There shall be no incomplete joint penetration in groove welds except as permitted for partial joint penetration welds.
3. The Test Supervisor shall examine the weld for acceptable appearance, and shall be satisfied that the welder is skilled in using the process and procedure specified for the text.
4. Undercut shall not exceed the lesser of 10% of the base metal thickness or 1/32 in. (0.8 mm).
5. Where visual examination is the only criterion for acceptance, all weld passes are subject to visual examination, at the discretion of the Test Supervisor.
6. The frequency of porosity shall not exceed one in each 4 in. (100 mm) of weld length and the maximum diameter shall not exceed 3/32 in. (2.4 mm)
7. Welds shall be free from overlap.

Sketches:

See Figure 2.49.

Complete a copy of the "Student Welding Report" listed in Appendix I or provided by your instructor.

PRACTICE 2-25

This practice parallels an AWS D1.1 structural steel limited thickness welder performance qualification test.

Module 1
Key Indicator 1, 2, 3, 4

Module 2
Key Indicator 1, 2, 3, 4, 7

Module 3
Key Indicator 1, 2, 3

Module 5
Short-circuit Transfer
Key Indicator 3, 4, 5, 6

Module 9
Key Indicator 1, 2

Gas Metal Arc Welding—Short-Circuit Metal Transfer (GMAW-S) Limited Thickness Welder Performance Qualification Test Plate for 2G, 3G, and 4G Positions without Backing

Welding Procedure Specification (WPS)

Welding Procedure Specification No.: Practice 2-25 Date:

Title:

Welding GMAW-S of plate to plate.

Scope:

This procedure is applicable for V-groove, bevel, or single-bevel welds within the range of 1/8 in. (3.2 mm) through 3/4 in. (19 mm).

Welding may be performed in the following positions: all.

Base Metal:

The base metal shall conform to carbon steel M-1, P-1, and S-1 Group 1 or 2.

Backing material specification: none.

Filler Metal:

The filler metal shall conform to AWS specification no. E70S-X from AWS specification A5.18. This filler metal falls into F-number F-6 and A-number A-1.

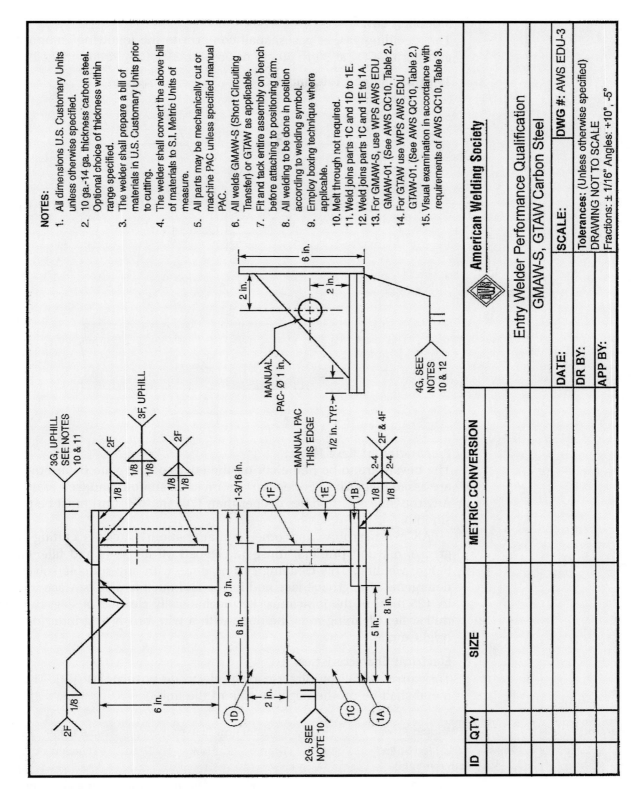

Figure 2.49 GMAW-S workmanship qualification test
Source: Courtesy of the American Welding Society

Shielding Gas:

The shielding gas, or gases, shall conform to the following compositions and purity: CO_2 at 30 to 50 cfh or 75% Ar/25% CO_2 at 30 to 50 cfh.

Joint Design and Tolerances:

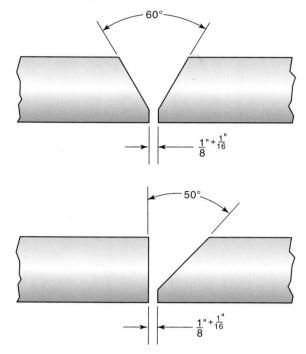

Preparation of Base Metal:

The bevels are to be flame cut on the edges of the plate before the parts are assembled. The beveled surface must be smooth and free of notches. Any roughness or notches deeper than 1/64 in. (0.4 mm) must be ground smooth.

All hydrocarbons and other contaminations, such as cutting fluids, grease, oil, and primers, must be cleaned off all parts and filler metals before welding. This cleaning can be done with any suitable solvents or detergents. The groove face and inside and outside plate surface within 1 in. (25 mm) of the joint must be mechanically cleaned of slag, rust, and mill scale. Cleaning must be done with a wire brush or grinder down to bright metal.

Electrical Characteristics:

The current shall be direct-current electrode positive (DCEP). The base metal shall be on the negative side of the line.

Electrode		Welding Power			Shielding Gas		Base Metal	
Type	Size	Amps	Wire-feed Speed, ipm (cm/min)	Volts	Type	Flow	Type	Thickness
E70S-X	0.035 in. (0.9 mm)	90 to 120	180 to 300 (457 to 762)	15 to 19	CO_2 or 75% Ar/ CO_2 25%	30 to 50	Low-carbon steel	1/4 in. to 1/2 in. (6 mm to 13 mm)
E70S-X	0.045 in. (1.2 mm)	130 to 200	125 to 200 (318 to 508)	17 to 20	CO_2 or 75% Ar/ CO_2 25%	30 to 50	Low-carbon steel	1/4 in. to 1/2 in. (6 mm to 13 mm)

Preheat:
The parts must be heated to a temperature higher than 50°F (10°C) before any welding is started.

Backing Gas:
N/A

Safety:
Proper protective clothing and equipment must be used. The area must be free of all hazards that may affect the welder or others in the area. The welding machine, welding leads, work clamp, electrode holder, and other equipment must be in safe working order.

Welding Technique:
Using a 1/2-in. (13-mm) or larger gas nozzle for all welding, first tack weld the plates together according to the drawing. There should be about a 1/8-in. (1.6-mm) root gap between the plates with V-grooved or beveled edges and 1/16-in. root faces. Use the E70S-X filler wire to make a root pass to fuse the plates together. Clean any silicon slag from the root pass, being sure to remove any trapped silicon slag along the sides of the weld.

Using the E70S-X filler wire, make a series of stringer or weave filler welds, no thicker than 1/4 in. (6.4 mm), in the groove until the joint is filled. Note: The horizontal (2G) weldment should be made with stringer beads only.

Interpass Temperature:
The plate should not be heated to a temperature higher than 350°F (175°C) during the welding process. After each weld pass is completed, allow it to cool but never to a temperature below 50°F (10°C). The weldment must not be quenched in water.

Cleaning:
Any slag must be cleaned off between passes. The weld beads may be cleaned by a hand wire brush, a hand chipping, a punch and hammer, or a needle-scaler. All weld cleaning must be performed with the test plate in the welding position.

Visual Inspection:*
Visually inspect the weld for uniformity and discontinuities.
1. There shall be no cracks, no incomplete fusion.
2. There shall be no incomplete joint penetration in groove welds except as permitted for partial joint penetration welds.
3. The Test Supervisor shall examine the weld for acceptable appearance, and shall be satisfied that the welder is skilled in using the process and procedure specified for the text.
4. Undercut shall not exceed the lesser of 10% of the base metal thickness or 1/32 in. (0.8 mm)
5. Where visual examination is the only criterion for acceptance, all weld passes are subject to visual examination, at the discretion of the Test Supervisor.

*From Table 3, AWS SENSE QC-10 2004 (Courtesy of the American Welding Society)

6. The frequency of porosity shall not exceed one in each 4 in. (100 mm) of weld length and the maximum diameter shall not exceed 3/32 in. (2.4 mm).

7. Welds shall be free from overlap.

Bend Test:

The weld is to be mechanically tested only after it has passed the visual inspection. Be sure that the test specimens are properly marked to identify the welder, the position, and the process.

Specimen Preparation:

For 3/8-in. test plates, two specimens are to be located in accordance with the requirements of the figure below left. One is to be prepared for a "transverse face bend," and the other is to be prepared for a "transverse root bend."

Transverse face bend. The weld is perpendicular to the longitudinal axis of the specimen and is bent so that the weld face becomes the tension surface of the specimen. Transverse face-bend specimens shall comply with the requirements of the figure below bottom.

Transverse root bend. The weld is perpendicular to the longitudinal axis of the specimen and is bent so that the weld root becomes the tension surface of the specimen. Transverse face-bend specimens shall comply with the requirements of the figure below bottom.

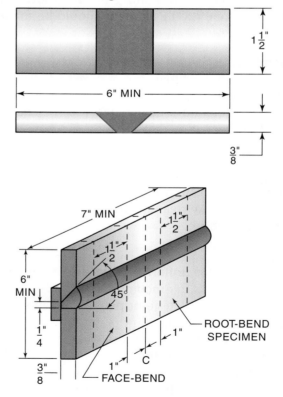

Acceptance Criteria for Face and Root Bends:*

For acceptance, the convex surface of the face- and root-bend specimens shall meet both of the following requirements:

*From Table 4, AWS-SENSE QC-10:2004 (Courtesy of the American Welding Society)

1. No single indication shall exceed 1/8 in. (3.2 mm), measured in any direction on the surface.
2. The sum of the greatest dimensions of all indications on the surface, which exceed 1/32 in. (0.8 mm) but are less than or equal to 1/8 in. (3.2 mm), shall not exceed 3/8 in. (9.6 mm).

Cracks occurring at the corner of the specimens shall not be considered unless there is definite evidence that they result from slag inclusion or other internal discontinuities.

Complete a copy of the "Student Welding Report" listed in Appendix I or provided by your instructor.

PRACTICE 2-26

Gas Metal Arc Welding Spray Transfer (GMAW) Workmanship Sample

Welding Procedure Specification (WPS)

Welding Procedure Specification No.: Practice 2-26. Date:

Module 1
Key Indicator 1, 2, 3, 4

Module 2
Key Indicator 1, 2, 3, 4, 7

Module 3
Key Indicator 1, 2, 3

Module 5
Spray Transfer
Key Indicator 8, 9, 10, 11, 12

Module 9
Key Indicator 1, 2

Title:

Welding GMAW of plate to plate.

Scope:

This procedure is applicable for V-groove and fillet welds within the range of 1/8 in. (3.2 mm) through 1-1/2 in. (38 mm).

Welding may be performed in the following positions: 1G, 1F, 2F.

Base Metal:

The base metal shall conform to carbon steel M-1, P-1, and S-1, Group 1 or 2.

Backing material specification: none.

Filler Metal:

The 0.035 to 0.045 dia. filler metal shall conform to AWS specification no. E70S-X from AWS specification A5.18. This filler metal falls into F-number F-6 and A-number A-1.

Shielding Gas:

The shielding gas, or gases, shall conform to the following compositions: 98% Ar/2% O_2 or 90% Ar/10% CO_2.

Preparation of Base Metal:

The bevels are to be flame cut on the edges of the plate before the parts are assembled. The beveled surface must be smooth and free of notches. Any roughness or notches deeper than 1/64 in. (0.4 mm) must be ground smooth.

All hydrocarbons and other contaminations, such as cutting fluids, grease, oil, and primers, must be cleaned off all parts and filler metals before welding. This cleaning can be done with any suitable solvents or detergents. The groove face and inside and outside plate surface within 1 in. (25 mm) of the joint must be mechanically cleaned of slag, rust, and mill scale. Cleaning must be done with a wire brush or grinder down to bright metal.

Electrical Characteristics:

The current shall be direct-current electrode positive (DCEP). The base metal shall be on the negative side of the line.

Electrode			Welding Power			Shielding Gas		Base Metal	
Type	Size	Amps	Wire-feed Speed, ipm (cm/min)	Volts	Type	Flow	Type	Thickness
E70S-X	0.035 in. (0.9 mm)	180 to 230	400 to 550 (1016 to 1397)	25 to 27	Ar plus 2% O_2 or 90% Ar/ 10% CO_2	30 to 50	Low-carbon steel	1/4 in. to 1/2 in. (6 mm to 13 mm)
E70S-X	0.045 in. (1.2 mm)	260 to 340	300 to 500 (762 to 1270)	25 to 30	Ar plus 2 O_2 or 90% Ar/ 10% CO_2	30 to 50	Low-carbon steel	1/4 in. to 1/2 in. (6 mm to 13 mm)

Preheat:

The parts must be heated to a temperature higher than 50°F (10°C) before any welding is started.

Backing Gas:

N/A

Safety:

Proper protective clothing and equipment must be used. The area must be free of all hazards that may affect the welder or others in the area. The welding machine, welding leads, work clamp, electrode holder, and other equipment must be in safe working order.

Welding Technique:

Using a 3/4-in. (19-mm) or larger gas nozzle for all welding, first tack weld the plates together according to the drawing. There should be about a 1/16-in. (1.6-mm) root gap between the plates with V-grooved or beveled edges. Use the E70S-X arc welding electrodes to make a root pass to fuse the plates together. Clean any silicon slag from the root pass, being sure to remove any trapped silicon slag along the sides of the weld.

Using the E70S-X arc welding electrodes, make a series of stringer or weave filler welds, no thicker than 1/4 in. (6.4 mm), in the groove until the joint is filled. The 1/4-in. (6.4-mm) fillet welds are to be made with one pass.

Interpass Temperature:

The plate should not be heated to a temperature higher than 350°F (175°C) during the welding process. After each weld pass is completed, allow it to cool but never to a temperature below 50°F (10°C). The weldment must not be quenched in water.

Cleaning:

Any slag must be cleaned off between passes. The weld beads may be cleaned by a hand wire brush, a hand chipping, a punch and hammer, or a needle-scaler. All weld cleaning must be performed with the test plate in the welding position.

Inspection:*

Visually inspect the weld for uniformity and discontinuities.
1. There shall be no cracks, no incomplete fusion.
2. There shall be no incomplete joint penetration in groove welds except as permitted for partial joint penetration welds.
3. The Test Supervisor shall examine the weld for acceptable appearance, and shall be satisfied that the welder is skilled in using the process and procedure specified for the text.

*From Table 3, AWS SENSE QC-10:2004 (Courtesy of the American Welding Society)

4. Undercut shall not exceed the lesser of 10% of the base metal thickness or 1/32 in. (0.8 mm)

5. Where visual examination is the only criterion for acceptance, all weld passes are subject to visual examination, at the discretion of the Test Supervisor.

6. The frequency of porosity shall not exceed one in each 4 in. (100 mm) of weld length and the maximum diameter shall not exceed 3/32 in. (2.4 mm).

7. Welds shall be free from overlap.

Sketches:
See Figure 2.50.

Complete a copy of the "Student Welding Report" listed in Appendix I or provided by your instructor.

PRACTICE 2-27

Gas Metal Arc Welding (GMAW) Spray Transfer Limited Thickness Welder Performance Qualification Test Plate for 1-G, Position, with Backing

Welding Procedure Specification (WPS)

Welding Procedure Specification No.: Practice (2-27). Date:

Title:
Welding GMAW of plate to plate.

Scope:
This procedure is applicable for V-groove and fillet welds within the range of 1/8 in. (3.2 mm) through 1-1/2 in. (38 mm).

Welding may be performed in the following positions: 1G, 2F.

Base Metal:
The base metal shall conform to carbon steel M-1, P-1, and S-1, Group 1 or 2.

Backing material specification: none.

Filler Metal:
The filler metal shall conform to AWS specification no. 0.035 to 0.045 dia. E70S-X from AWS specification A5.18. This filler metal falls into F-number F-6 and A-number A-1.

Shielding Gas:
The shielding gas, or gases, shall conform to the following compositions and purity: 98% Ar/2% O_2 or 90% Ar/10% CO_2

Joint Design and Tolerances:

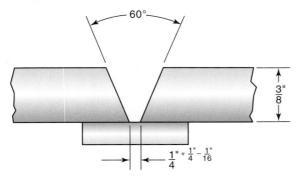

Module 1
Key Indicator 1, 2, 3, 4

Module 2
Key Indicator 1, 2, 3, 4, 7

Module 3
Key Indicator 1, 2, 3

Module 5
Spray Transfer
Key Indicator 8, 9, 11

Module 9
Key Indicator 1, 2

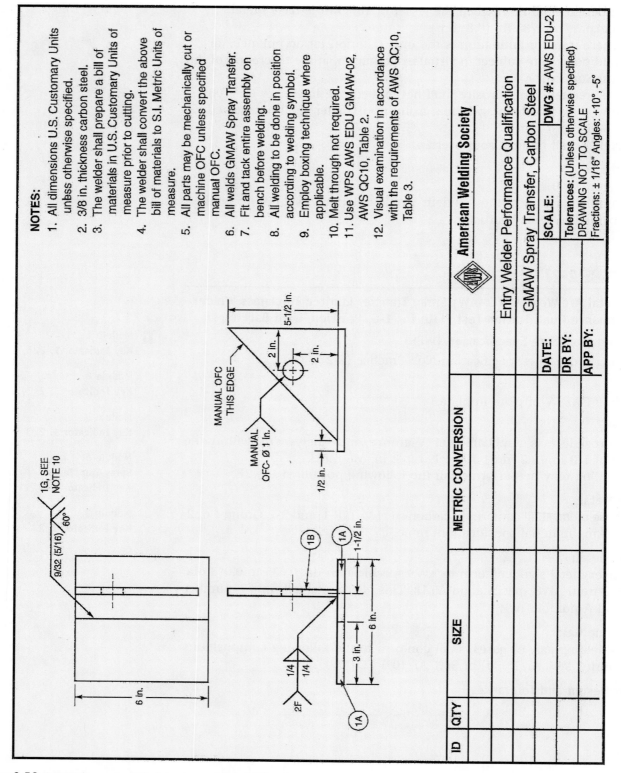

Figure 2.50 GMAW (spray transfer) workmanship qualification test
Source: Courtesy of the American Welding Society

Preparation of Base Metal:

The bevels are to be flame cut on the edges of the plate before the parts are assembled. The beveled surface must be smooth and free of notches. Any roughness or notches deeper than 1/64 in. (0.4 mm) must be ground smooth.

All hydrocarbons and other contaminations, such as cutting fluids, grease, oil, and primers, must be cleaned off all parts and filler metals before welding. This cleaning can be done with any suitable solvents or detergents. The groove face and inside and outside plate surface within 1 in. (25 mm) of the joint must be mechanically cleaned of slag, rust, and mill scale. Cleaning must be done with a wire brush or grinder down to bright metal.

Electrical Characteristics:

The current shall be direct-current electrode positive (DCEP). The base metal shall be on the negative side of the line.

Electrode		Welding Power			Shielding Gas		Base Metal	
Type	Size	Amps	Wire-feed Speed ipm (cm/min)	Volts	Type	Flow	Type	Thickness
E70S-X	0.035 in. (0.9 mm)	180 to 230	400 to 550 (1016 to 1397)	25 to 28	Ar plus 2% O_2 or 90% Ar/10% CO_2	30 to 50	Low-carbon steel	1/4 in. to 1/2 in. (6 mm to 13 mm)
E70S-X	0.045 in. (1.2 mm)	260 to 340	300 to 500 (762 to 1270)	25 to 30	Ar plus 2 O_2 or 90% Ar/10% CO_2	30 to 50	Low-carbon steel	1/4 in. to 1/2 in. (6 mm to 13 mm)

Preheat:

The parts must be heated to a temperature higher than 50°F (10°C) before any welding is started.

Backing Gas:

N/A

Safety:

Proper protective clothing and equipment must be used. The area must be free of all hazards that may affect the welder or others in the area. The welding machine, welding leads, work clamp, electrode holder, and other equipment must be in safe working order.

Welding Technique:

Using a 3/4-in. (19-mm) or larger gas nozzle for all welding, first tack weld the plates and backing strip together according to the drawing below. There should be about a 1/4-in. (1.6-mm) root gap between the plates with V-grooved or beveled edges. Use the E70S-X arc welding electrodes to make a root pass to fuse the plates together. Clean any silicon slag from the root pass, being sure to remove any trapped silicon slag along the sides of the weld.

Using the E70S-X arc welding electrodes, make a series of stringer filler welds, no thicker than 1/4 in. (6.4 mm), in the groove until the joint is filled.

Interpass Temperature:

The plate should not be heated to a temperature higher than 350°F (175°C) during the welding process. After each weld pass is completed, allow it to cool but never to a temperature below 50°F (10°C). The weldment must not be quenched in water.

Cleaning:

Any slag must be cleaned off between passes. The weld beads may be cleaned by a hand wire brush, a hand chipping, a punch and hammer, or a needle-scaler. All weld cleaning must be performed with the test plate in the welding position.

Visual Inspection:*

Visually inspect the weld for uniformity and discontinuities.
1. There shall be no cracks, no incomplete fusion.
2. There shall be no incomplete joint penetration in groove welds except as permitted for partial joint penetration welds.
3. The Test Supervisor shall examine the weld for acceptable appearance, and shall be satisfied that the welder is skilled in using the process and procedure specified for the text.
4. Undercut shall not exceed the lesser of 10% of the base metal thickness or 1/32 in. (0.8 mm).
5. Where visual examination is the only criterion for acceptance, all weld passes are subject to visual examination, at the discretion of the Test Supervisor.
6. The frequency of porosity shall not exceed one in each 4 in. (100 mm) of weld length and the maximum diameter shall not exceed 3/32 in. (2.4 mm).
7. Welds shall be free from overlap.

Sketches:

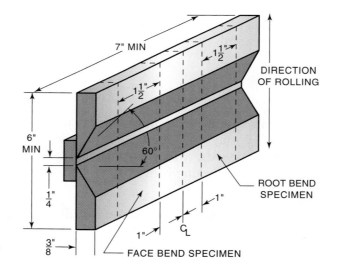

Bend-Test:

The weld is to be mechanically tested only after it has passed the visual inspection. Be sure that the test specimens are properly marked to identify the welder, the position, and the process.

Specimen Preparation

For 3/8-in. test plates, two specimens are to be located in accordance with the requirements below. One is to be prepared for a transverse face bend, and the other is to be prepared for a transverse root bend.

*From Table 3, AWS SENSE QC-10:2004 (Courtesy of the American Welding Society)

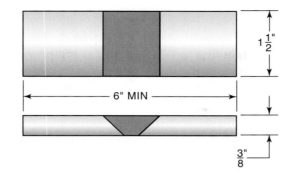

Transverse face bend. The weld is perpendicular to the longitudinal axis of the specimen and is bent so that the weld face becomes the tension surface of the specimen. Transverse face bend specimens shall comply with the requirements below.

Transverse root bend. The weld is perpendicular to the longitudinal axis of the specimen and is bent so that the weld root becomes the tension surface of the specimen. Transverse root bend specimens shall comply with the requirements below.

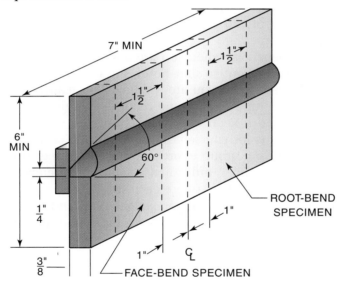

Acceptance Criteria for Face and Root Bends*

For acceptance, the convex surface of the face and root bend specimens shall meet both of the following requirements:

1. No single indication shall exceed 1/8 in. (3.2 mm), measured in any direction on the surface.
2. The sum of the greatest dimensions of all indications on the surface, which exceed 1/32 in. (0.8 mm), but are less than or equal to 1/8 in. (3.2 mm), shall not exceed 3/8 in. (9.6 mm).

Cracks occurring at the corner of the specimens shall not be considered unless there is definite evidence that they result from slag or inclusions or other internal discontinuities.

Complete a copy of the "Student Welding Report" listed in Appendix I or provided by your instructor.

*From Table 4, AWS SENSE QC-10:2004 (Courtesy of the American Welding Society)

SUMMARY

Slight changes in welding gun angle and electrode extension can make significant differences in the quality of the weld produced. As a new welder you might find it difficult to tell the effect of these changes if they are slight. Therefore, as you start to learn this process it is a good idea to make more radical changes so it is easier for you to see their effects on the weld. Later as you develop your skills you can use these slight changes to aid in controlling the weld's quality and appearance as it progresses along the joint. Small adjustments in your welding technique are required to compensate for slight changes that occur along a welding joint, such as joint gap and the increasing temperature of the base metal.

Variations in conditions can significantly affect welding setup for the GMA process. Before starting an actual weld in the field you should practice to test your setup. Practice on scrap metal of a similar thickness and type of metal to be welded. A practice weld before you begin welding can significantly increase the chances that your weld will meet standards or specifications. Making these sample or test welds is more important when you are welding in the field, since welds outside the shop are more difficult to control and anticipate. Think of this much as an athlete warms up before competing.

You will find it beneficial when you are initially setting up your welder to have someone assist you, so that he or she can make changes in the welding machine's settings as you are welding. This teamwork can significantly increase your setup accuracy and reduce setup time. Later on in the field, having developed a keen eye for watching the weld, you can then make these adjustments for yourself more rapidly and accurately. Working with another student in a group effort like this will also give you a better understanding of how other individuals' setup preferences affect their welds. Welding is an art, and therefore each welder may have slight differences in preference for voltage, amperage, gas flow, and other setup variables. This gives you an opportunity to learn more from others.

REVIEW

1. What items make up a basic semiautomatic welding system?
2. What must be done to the shielding gas cylinder before the valve protection cap is removed?
3. Why is the shielding gas valve "cracked" before the flowmeter regulator is attached?
4. What causes the electrode to bird-nest?
5. Why must all fittings and connections be tight?
6. What parts should be activated by depressing the gun switch?
7. What benefit does a welding wire's cast provide?
8. What can be done to determine the location of a problem that stops the wire from being successfully fed through the conduit?
9. What are the advantages of using a feed roller pressure that is as light as possible?

10. Why should the feed roller drag prevent the spool from coasting to a stop when the feed stops?
11. Why must you always wind the wire tightly into a ball or cut it into short lengths before discarding it in the proper waste container?
12. Why would the flowmeter ball float at different heights with different shielding gases if the shielding gases are flowing at the same rate?
13. Using Table 2.1, determine the amperage if 400 in. (10.2 m) of 0.45-in. (1.2-mm) steel wire is fed in one minute.
14. How is the amperage adjusted on a GMA welder?
15. What happens to the weld as the electrode extension is lengthened?
16. What is the effect on the weld of changing the welding angle from a dragging to a pushing angle?
17. What are the advantages of adding oxygen or CO_2 to argon for welds on steel?
18. What are the advantages of using CO_2 for making GMA welds on steel?
19. What is mill scale?
20. What type of porosity is most often caused by mill scale?
21. What should the welder watch if the view of the weld is obstructed by the shielding gas nozzle?
22. When you are making a vertical weld and it appears that the weld metal is going to drip over the shelf, what should you do?
23. What are the advantages of making vertical down welds?
24. How can small weld beads be maintained during overhead welds?
25. How can spatter be controlled on the nozzle when making overhead welds?
26. How should the electrode be manipulated for the deepest penetration when using the pulsed-arc metal transfer process?

CHAPTER

3

Flux Cored Arc Welding Equipment, Setup, and Operation

OBJECTIVES

After completing this chapter, the student should be able to

- describe the flux cored arc (FCA) welding process
- list the equipment required for an FCA welding workstation
- list five advantages of FCA welding, and explain four of its limitations
- tell how electrodes are manufactured and explain the purpose of the electrode cast and helix
- list four things flux can provide to the weld and how fluxes are classified
- explain what each of the digits in a standard FCAW electrode identification number mean
- describe the proper care and handling of FCAW electrodes
- list two common shielding gases used in FCAW, and contrast their benefits related to cost, productivity, and quality
- list three differences in an FCA weld when the gun angle is changed
- identify the two modes of metal transfer and contrast them in regard to application and quality
- list four effects that electrode extension has on FCA welding
- list three things that can cause weld porosity and how it can be prevented

KEY TERMS

air-cooled

coils

deoxidizers

dual shield

flux cored arc welding (FCAW)

lime-based flux

rutile-based flux

self-shielding

slag

smoke extraction nozzles

spools

water-cooled

AWS SENSE EG2.0

Key Indicators Addressed in this Chapter:

Module 6: Flux Cored Arc Welding (FCAW-G, FCAW-S)

Key Indicator 1: Performs safety inspections of FCAW-G/GM, FCAW-S equipment and accessories

Key Indicator 2: Makes minor external repairs to FCAW-G/GM, FCAW-S equipment and accessories

Key Indicator 3: Sets up for FCAW-G/GM, FCAW-S operations on carbon steel

INTRODUCTION

Flux cored arc welding (FCAW) is a fusion welding process in which weld heating is produced from an arc between the work and a continuously fed filler metal electrode. Atmospheric shielding is provided completely or in part by the flux sealed within the tubular electrode, Figure 3.1. Extra shielding may or may not be supplied through a nozzle in the same way as in GMAW.

Although the process was introduced in the early 1950s, it represented less than 5% of the total amount of welding done in 1965. In 2005, it passed the 50% mark and is still rising. The rapid rise in the use of FCAW has been due to a number of factors. Improvements in the fluxes, smaller electrode diameters, increased reliability of the equipment, better electrode feed systems, and improved guns have all led to the increased usage. Guns equipped with **smoke extraction nozzles** and electronic controls are the latest in a long line of improvements to this process, Figure 3.2.

PRINCIPLES OF OPERATION

FCA welding is similar in a number of ways to the operation of GMA welding, Figure 3.3. Both processes use a constant-potential (CP) or constant-voltage (CV) power supply. Constant potential and voltage are terms that have the same meaning. CP power supplies provide a controlled voltage (potential) to the welding electrode. The amperage (current) varies with the speed that the electrode is being fed into the molten weld pool. Just as in GMA welding, higher electrode feed speeds produce higher currents and slower feed speeds result in lower currents, assuming all other conditions remain constant.

The effects on the weld of electrode extension, gun angle, welding direction, travel speed, and other welder manipulations are similar to those experienced in GMA welding. As in GMA welding, having a correctly set welder does not ensure a good weld. The skill of the welder is an important factor in producing high-quality welds.

The flux inside the electrode protects the molten weld pool from the atmosphere, improves strength through chemical reactions and alloys, and improves the weld shape.

Atmospheric contamination of molten weld metal occurs as it travels across the arc gap and within the pool before it solidifies. The major atmospheric contaminations come from oxygen and nitrogen, the major elements in air. The addition of fluxing and gas-forming elements to the core electrode reduces or eliminates their effects.

Improved strength and other physical or corrosion-resistant properties of the finished weld are improved by the flux. Small additions of alloying elements, deoxidizers, and gas-forming and slag agents all can improve the desired weld properties. Carbon, chromium, and vanadium can be added to improve hardness, strength, creep resistance, and corrosion resistance. Aluminum, silicon, and titanium all help remove oxides and/or nitrides in the weld. Potassium, sodium, and zirconium are added to the flux and form a slag.

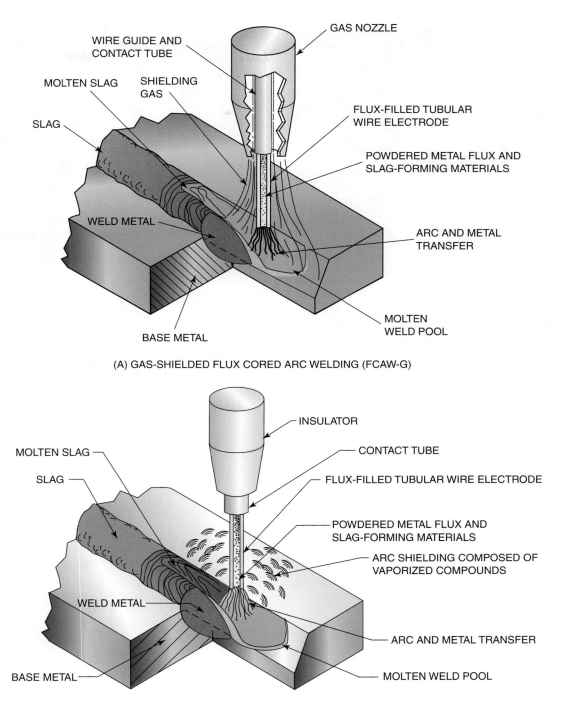

WIRE GUIDE AND
CONTACT TUBE

GAS NOZZLE

MOLTEN SLAG

SHIELDING
GAS

SLAG

FLUX-FILLED TUBULAR
WIRE ELECTRODE

POWDERED METAL FLUX AND
SLAG-FORMING MATERIALS

WELD METAL

ARC AND METAL
TRANSFER

MOLTEN
WELD POOL

BASE METAL

(A) GAS-SHIELDED FLUX CORED ARC WELDING (FCAW-G)

INSULATOR

MOLTEN SLAG

CONTACT TUBE

SLAG

FLUX-FILLED TUBULAR WIRE ELECTRODE

POWDERED METAL FLUX AND
SLAG-FORMING MATERIALS

ARC SHIELDING COMPOSED OF
VAPORIZED COMPOUNDS

WELD METAL

ARC AND METAL TRANSFER

BASE METAL

MOLTEN WELD POOL

(B) SELF-SHIELDED FLUX CORED ARC WELDING (FCAW-S)

Figure 3.1 Two types of flux cored arc welding
FCA welding may have extra shielding provided by a gas nozzle (A), or be self-shielding only (B).
Source: Courtesy of the American Welding Society

A discussion of weld metal additives and flux elements and their effects on the weld can be found later in this chapter.

The flux core additives that serve as deoxidizers, gas formers, and slag formers either protect the molten weld pool or help to remove impurities from the base metal. Deoxidizers may convert small amounts of surface

(A)   (B)

(C)   (D)

Figure 3.2 Smoke extraction
(A) FCA welding without smoke extraction and (B) with smoke extraction. (C) Typical FCAW smoke extraction gun. (D) Typical smoke
exhaust system.
Source: Courtesy of Lincoln Electric Company

oxides like mill scale back into pure metal. They work much like the elements used to refine iron ore into steel.

Gas formers rapidly expand and push the surrounding air away from the molten weld pool. If oxygen in the air were to come in contact with the molten weld metal, the weld metal would quickly oxidize. Sometimes this can be seen at the end of a weld when the molten weld metal erupts in a shower of tiny sparks.

Figure 3.3 Large-capacity wire-feed unit used with FCAW or GMAW
Source: Courtesy of Lincoln Electric Company

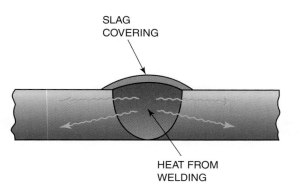

SLAG
COVERING

HEAT FROM
WELDING

Figure 3.4 Slag blanketing the weld
The slag covering keeps the welding heat from escaping quickly, thus slowing the cooling rate.

The slag covering of the weld is useful for several reasons. **Slag** is a nonmetallic product resulting from the mutual dissolution of the flux and nonmetallic impurities in the base metal. Slag helps the weld by protecting the hot metal from the effects of the atmosphere, controlling the bead shape by serving as a dam or mold, and serving as a blanket to slow the weld's cooling rate, which improves its physical properties, Figure 3.4.

EQUIPMENT
Power Supply

Module 6
Key Indicator 1, 2, 3

The FCA welding power supply is the same type that is required for GMAW, called constant-potential, constant-voltage (CP, CV). The words *potential* and *voltage* have the same electrical meaning and are used inter-changeably. FCAW machines can be much more powerful than GMAW machines and are available with up to 1500 amperes of welding power.

Guns

FCA welding guns are available as **water-cooled** or **air-cooled**, Figure 3.5. Although most of the FCA welding guns that you will find in schools are air-cooled, our industry often needs water-cooled guns because of the

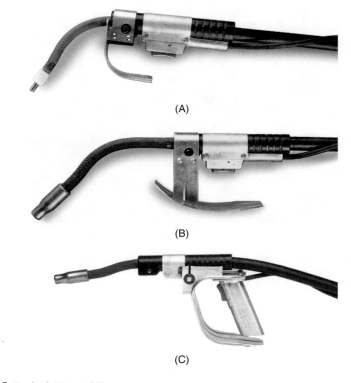

Figure 3.5 Typical FCA welding guns
(A) 350 ampere rating self-shielding, (B) 450 ampere rating gas-shielding, and (C) 600 ampere rating gas-shielding.
Source: Courtesy of Lincoln Electric Company

higher heat caused by longer welds made at higher currents. The water-cooled FCA welding gun is more efficient than an air-cooled gun at removing waste heat. The air-cooled gun is more portable because it has fewer hoses, and it may be made lighter so it is easier to manipulate than the water-cooled gun.

Also, the water-cooled gun requires a water reservoir or another system to give the needed cooling. There are two major ways that water can be supplied to the gun for cooling. Cooling water can be supplied directly from the building's water system, or it can be supplied from a recirculation system.

Cooling water supplied directly from the building's water system is usually dumped into a wastewater drain once it has passed through the gun. When this type of system is used, a pressure regulator must be installed to prevent pressures that are too high from damaging the hoses. Water pressures higher than 35 psi (241 kg/mm^2) may cause the water hoses to burst. Check valves must also be installed in the supply line to prevent contaminated water from being drawn back into the water supply. Some cities and states have laws that restrict the use of open systems because of the need for water conservation. Check with your city or state for any restrictions before installing an open water-cooling system.

Recirculating cooling water systems eliminate any of the problems associated with open systems. Chemicals may be added to the water in recirculating systems to prevent freezing, to aid in pump lubrication, and to prevent algae growth. Only manufacturer-approved additives should be used in a recirculation system. Read all of the manufacturer's safety and data sheets before using these chemicals.

Smoke Extraction Nozzles

Because of the large quantity of smoke that can be generated during FCA welding, systems for smoke extraction that fit on the gun have been designed, Figure 3.2B. These systems use a vacuum to pull the smoke back into a specially designed smoke extraction nozzle on the welding gun. The disadvantage of this slightly heavier gun is offset by the system's advantages. The advantages of the system are as follows:

- Cleaner air for the welder to breathe because the smoke is removed before it rises to the welder's face.
- Reduced heating and cooling cost because the smoke is concentrated, so less shop air must be removed with the smoke.

Electrode Feed

Electrode feed systems are similar to those used for GMAW; in fact many feed systems are designed with dual feeders so that solid wire and flux core may be run in sequence. The major difference is that the more robust FCAW feeders are designed to use large-diameter wire and most often have two sets of feed rollers. The two sets of rollers help reduce the drive pressure on the electrode. Excessive pressure can distort the electrode wire diameter, which can allow some flux to be dropped inside the electrode guide tube.

ADVANTAGES

FCA welding offers the welding industry a number of important advantages.

High Deposition Rate

High rates of depositing weld metal are possible. FCA welding deposition rates of more than 25 lb/hr (11 kg/hr) of weld metal are possible. This compares to about 10 lb/hr (5 kg/hr) for shielded metal arc (SMA) welding using a very large-diameter electrode of 1/4 in. (6 mm).

Minimum Electrode Waste

The FCA method makes efficient use of filler metal; from 75% to 90% of the weight of the FCAW electrode is metal, the remainder being flux. SMAW electrodes have a maximum of 75% filler metal; some SMAW electrodes have much less. Also, a stub must be left at the end of each SMA welding electrode. The stub will average 2 in. (51 mm) in length, resulting in a loss of 11% or more of the SMAW filler electrode purchased. FCA welding has no stub loss, so nearly 100% of the FCAW electrode purchased is used.

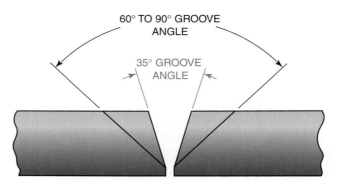

Figure 3.6 Narrower groove angle for FCAW
The narrower groove angle for FCAW compared to other welding processes saves on filler metal, welding time, and heat input into the part.

Narrow Groove Angle

Because of the deep penetration characteristic of FCAW, no edge-beveling preparation is required on some joints in metal up to 1/2 in. (13 mm) in thickness. When bevels are cut, the joint-included angle can be reduced to as small as 35°, Figure 3.6. The reduced groove angle results in a smaller-sized weld. This can save 50% of filler metal with about the same savings in time and weld power used.

Minimum Precleaning

The addition of **deoxidizers**, which combine with and remove harmful oxides on the base metal or its surface, and other fluxing agents permits high-quality welds to be made on plates with light surface oxides and mill scale. This eliminates most of the precleaning required before GMA welding can be performed. Often it is possible to make excellent welds on plates in the "as cut" condition; no cleanup is needed.

All-position Welding

Small-diameter electrode sizes in combination with special fluxes allow excellent welds in all positions. The slags produced assist in supporting the weld metal. This process is easy to use, and, when properly adjusted, it is much easier to use than other all-position arc welding processes.

Flexibility

Changes in power settings can permit welding to be done on thin-gauge sheet metals or thicker plates using the same electrode size. Multipass welds allow joining metals with no limit on thickness. This, too, is attainable with one size of electrode.

High Quality

Many codes permit welds to be made using FCAW. The addition of the flux gives the process the high level of reliability needed for welding on boilers, pressure vessels, and structural steel.

Excellent Control

The molten weld pool is more easily controlled with FCAW than with GMAW. The surface appearance is smooth and uniform even with less operator skill. Visibility is improved by removing the nozzle when using self-shielded electrodes.

LIMITATIONS

The main limitation of flux cored arc welding is that it is confined to ferrous metals and nickel-based alloys. Generally, all low- and medium-carbon steels and some low-alloy steels, cast irons, and a limited number of stainless steels are presently weldable using FCAW.

The equipment and electrodes used for the FCAW process are more expensive. However, the cost is quickly recoverable through higher productivity.

The removal of postweld slag requires another production step. The flux must be removed before the weldment is finished (painted) to prevent crevice corrosion.

With the increased welding output comes an increase in smoke and fume generation. The existing ventilation system in a shop might need to be increased to handle the added volume.

ELECTRODES

Methods of Manufacturing

The electrodes have flux tightly packed inside. One method used to make them is to first form a thin sheet of metal into a U-shape, Figure 3.7. A measured quantity of flux is poured into the U-shape before it is squeezed shut. The wire is then passed through a series of dies to size it and further compact the flux.

A second method of manufacturing the electrode is to start with a seamless tube. The tube is usually about 1 in. in diameter. One end of the tube is sealed, and the flux powder is poured into the open end. The tube is vibrated during the filling process to ensure that it fills completely. Once the tube is full, the open end is sealed. The tube is now sized using a series of dies, Figure 3.8.

In both these methods of manufacturing the electrode, the sheet and tube are made up of the desired alloy. Also in both cases, the flux is compacted inside the metal skin. This compacting helps make the electrode operate more smoothly and consistently.

Electrodes are available in sizes from 0.030 in. to 5/32 in. (0.8 mm to 3.9 mm) in diameter. Smaller-diameter electrodes are much more expensive per pound than the same type in a larger diameter due to the high cost of drawing and filling cored wires to small sizes. Larger-diameter electrodes produce such large welds they cannot be controlled in all positions. The most popular diameters range from 0.035 in. to 3/32 in. (0.9 mm to 2.3 mm).

The finished FCA filler metal is packaged in a number of forms for purchase by the end user, Figure 3.9. The AWS has a standard for the size of

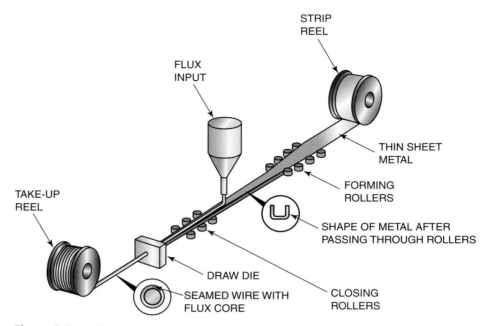

Figure 3.7 Putting the flux in the flux cored wire

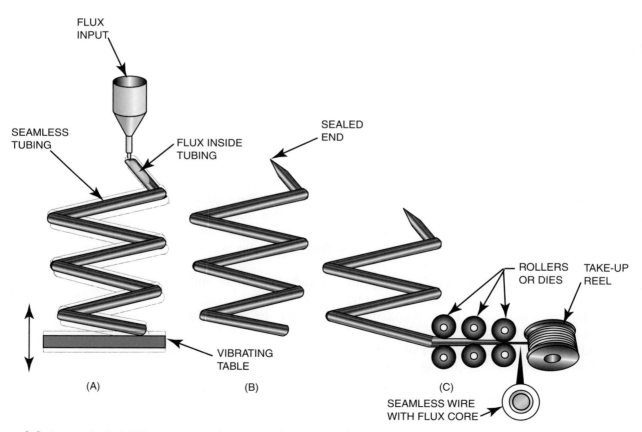

Figure 3.8 One method of filling seamless FCA welding filler metal with flux
The vibration helps compact the granular flux inside the tube.

Figure 3.9 A few packaged forms of FCA filler metal
FCAW filler metal weights are approximate. They will vary by alloy and manufacturer.
Source: Courtesy of Lincoln Electric Company

each of the package units. Although the dimensions of the packages are standard, the weight of filler wire is not standard. More of the smaller-diameter wire can fit into the same space compared with a larger-diameter wire, so a package of 0.030-in. (0.8-mm) wire weighs more than the same-sized package of 3/32-in. (2.3-mm) wire. The standard packing units for FCAW wires are **spools**, **coils**, reels, and drums, Table 3.1.

Spools are made of plastic or fiberboard and are disposable. They are completely self-contained and are available in approximate weights from

Table 3.1 Packaging Size Specification for Commonly Used FCA Filler Wire

Packaging	Outside Diameter	Width	Arbor (Hole) Diameter
Spools	4 in. (102 mm)	1-3/4 in. (44.5 mm)	5/8 in. (16 mm)
	8 in. (203 mm)	2-1/4 in. (57 mm)	2-1/16 in. (52.3 mm)
	12 in. (305 mm)	4 in. (102 mm)	2-1/16 in. (52.3 mm)
	14 in. (356 mm)	4 in. (102 mm)	2-1/16 in. (52.3 mm)
Reels	22 in. (559 mm)	12-1/2 in. (318 mm)	1-5/16 in. (33.3 mm)
	30 in. (762 mm)	16 in. (406 mm)	1-5/16 in. (33.3 mm)
Coils	16-1/4 in. (413 mm)	4 in. (102 mm)	12 in. (305 mm)

	Outside Diameter	Inside Diameter	Height
Drums	23 in. (584 mm)	16 in. (406 mm)	34 in. (864 mm)

1 lb up to around 50 lb (0.5 kg to 25 kg). The smaller spools, 4 in. and 8 in. (102-mm and 203 mm), weighing from 1 lb to 7 lb, are most often used for smaller production runs or for home/hobby use; 12-in. and 14-in. (305-mm and 356-mm) spools are often used in schools and welding fabrication shops.

Coils come wrapped and/or wire tied together. They are unmounted, so they must be supported on a frame on the wire feeder in order to be used. Coils are available in weights around 60 lb (27 kg). Because FCAW wires on coils do not have the expense of a disposable core, these wires cost a little less per pound, so they are more desirable for higher-production shops.

Reels are large wooden spools, and drums are shaped like barrels. Both reels and drums are used for high-production jobs. Both can contain approximately 300 lb to 1000 lb (136 kg to 454 kg) of FCAW wire. Because of their size, they are used primarily at fixed welding stations. Such stations are often associated with some form of automation, such as turntables or robotics.

Electrode Cast and Helix

To see the cast and helix of a wire, feed out 10 ft of wire electrode and cut it off. Lay it on the floor and observe that it forms a circle. The diameter of the circle is known as the cast of the wire, Figure 3.10.

Note that the wire electrode does not lay flat. One end is slightly higher than the other. This height is the helix of the wire.

The AWS has specifications for both cast and helix for all FCA welding wires.

The cast and helix cause the wire to rub on the inside of the contact tube, Figure 3.11. The slight bend in the electrode wire ensures a positive electrical contact between the contact tube and filler wire.

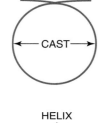

Figure 3.10 Method of measuring cast and helix of FCAW filler wire

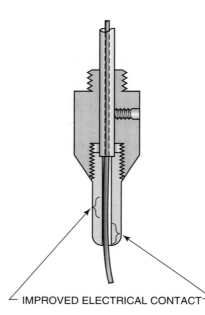

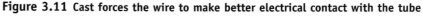

Figure 3.11 Cast forces the wire to make better electrical contact with the tube

FLUX

The fluxes used are mainly based on lime or rutile (a mineral compound consisting of titanium dioxide, usually with a little iron). The purpose of the fluxes is the same as in the shielded metal arc welding (SMAW) process. That is, they can provide all or part of the following to the weld:

- *Deoxidizers:* Oxygen that is present in the welding zone has two forms. It can exist as free oxygen from the atmosphere surrounding the weld. Oxygen can also exist as part of a compound such as an iron oxide or carbon dioxide (CO_2). In either case it can cause porosity in the weld if it is not removed or controlled. Chemicals are added that react to the presence of oxygen in either form and combine to form a harmless compound, Table 3.2. The new compound can become part of the slag that solidifies on top of the weld, or some of it may stay in the weld as very small inclusions. Both methods result in a weld with better mechanical properties because of lower porosity.
- *Slag formers:* Slag serves several vital functions for the weld. It can react with the molten weld metal chemically, and it can affect the weld bead physically. In the molten state it moves through the molten weld pool and acts as a magnet or sponge to chemically combine with impurities in the metal and remove them, Figure 3.12. Slags can be refractory, so that they become solid at a high temperature. As they solidify over the weld, they help it hold its shape and they slow its cooling rate.
- *Fluxing agents:* Molten weld metal tends to have a high surface tension, which prevents it from flowing outward toward the edges of the weld. This causes undercutting along the junction of the

Table 3.2 Deoxidizing Elements Added to Filler Wire (to Minimize Porosity in the Molten Weld Pool)

Deoxidizing Element		Strength
Aluminum (Al)		Very strong
Manganese (Mn)		Weak
Silicon (Si)		Weak
Titanium (Ti)		Very strong
Zirconium (Zr)		Very strong

Figure 3.12 Impurities being floated to the surface by slag

weld and the base metal. Fluxing agents make the weld more fluid and allow it to flow outward, filling the undercut.

- *Arc stabilizers:* Chemicals in the flux affect the arc resistance. As the resistance is lowered, the arc voltage drops and penetration is reduced. When the arc resistance is increased, the arc voltage increases and weld penetration is increased. Although the resistance within the ionized arc stream may change, the arc is more stable and easier to control. It also improves the metal transfer by reducing spatter caused by an erratic arc.

- *Alloying elements:* Because of the difference in the mechanical properties of metal that is formed by rolling or forging and metal that is melted to form a weld bead, the metallurgical requirements of the two also differ. Some elements change the weld's strength, ductility, hardness, brittleness, toughness, and corrosion resistance. Other alloying elements in the form of powder metal can be added to increase deposition rates.

- *Shielding gas:* As elements in the flux are heated by the arc, some of them vaporize and form voluminous gaseous clouds hundreds of times larger than their original volume. This rapidly expanding cloud forces the air around the weld zone away from the molten weld metal, Figure 3.13. Without the protection this process affords the molten metal, it would rapidly oxidize. Such oxidization would severely affect the weld's mechanical properties, rendering it unfit for service.

All FCAW fluxes are divided into two groups based on the acid or basic chemical reactivity of the slag. The AWS classifies T-1 as acid and T-5 as basic.

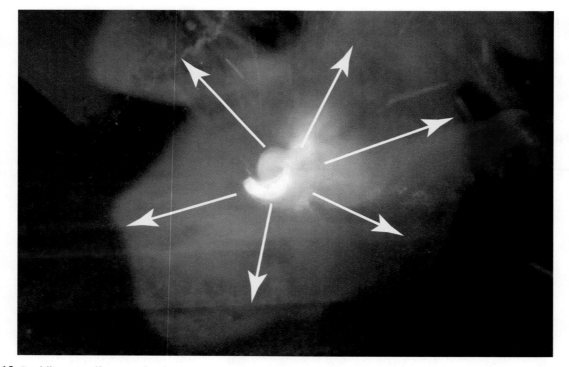

Figure 3.13 Rapidly expanding gas cloud
Source: Courtesy of Larry Jeffus

The **rutile-based flux** is acidic, T-1. It produces a smooth, stable arc and a refractory high-temperature slag for out-of-position welding. These electrodes produce a fine drop transfer, a relatively low fume, and an easily removed slag. The main limitation of the rutile fluxes is that their fluxing elements do not produce as high a quality deposit as do the T-5 systems.

The **lime-based flux** is basic, T-5. It is very good at removing certain impurities from the weld metal, but its low-melting-temperature slag is fluid, which makes it generally unsuitable for out-of-position welding. These electrodes produce a more globular transfer, more spatter, more fume, and a more adherent slag than do the T-1 systems. These characteristics are tolerated when it is necessary to deposit very tough weld metal and for welding materials having a low tolerance for hydrogen.

Some rutile-based electrodes allow the addition of a shielding gas. With the weld partially protected by the shielding gas, more elements can be added to the flux, which produces welds with the best of both flux systems, high-quality welds in all positions.

Some fluxes can be used on both single- and multiple-pass welds, and others are limited to single-pass welds only. Using a single-pass welding electrode for multipass welds may result in an excessive amount of manganese. The manganese is necessary to retain strength when making large, single-pass welds. However, with the lower dilution associated with multipass techniques, it can strengthen the weld metal too much and reduce its ductility. In some cases, small welds that deeply penetrate the base metal can help control this problem.

Table 3.3 lists the shielding and polarity for the flux classifications of mild steel FCAW electrodes. The letter *G* is used to indicate an unspecified classification. The *G* means that the electrode has not been classified by the American Welding Society. Often the exact composition of fluxes is kept as a manufacturer's trade secret. Therefore, only limited information about the electrode's composition will be given. The only information often supplied is current, type of shielding required, and some strength characteristics.

Table 3.3 Welding Characteristics of Seven Flux Classifications

Classification	Comments	Shielding Gas
T-1	Requires clean surfaces and produces little spatter. It can be used for single- and multiple-pass welds in all positions.	Carbon dioxide (CO_2) or argon/carbon dioxide mixes
T-2	Requires clean surfaces and produces little spatter. It can be used for single-pass welds in the flat (1G and 1F) and horizontal (2F) positions only.	Carbon dioxide (CO_2)
T-3	Used on thin-gauge steel for single-pass welds in the flat (1G and 1F) and horizontal (2F) positions only.	None
T-4	Low penetration and moderate tendency to crack for single- and multiple-pass welds in the flat (1G and 1F) and horizontal (2F) positions.	None
T-5	Low penetration and a thin, easily removed slag, used for single- and multiple-pass welds in the flat (1G and 1F) position only.	With or without carbon dioxide (CO_2)
T-6	Similar to T-5 without externally applied shielding gas.	None
T-G	The composition and classification of this electrode are not given in the preceding classes. It may be used for single- or multiple-pass welds.	With or without shielding

Table 3.4 Ferrite-forming Elements Used in FCA Welding Fluxes

Element	Reaction in Weld
Silicon (Si)	Ferrite former and deoxidizer
Chromium (Cr)	Ferrite and carbide former
Molybdenum (Mo)	Ferrite and carbide former
Columbium (Cb)	Strong ferrite former
Aluminum (Al)	Ferrite former and deoxidizer

As a result of the relatively rapid cooling of the weld metal, the weld may tend to become hard and brittle. This factor can be controlled by adding elements to the flux that affect the content of both the weld and the slag, Table 3.4. Ferrite is the softer, more ductile form of iron. The addition of ferrite-forming elements can control the hardness and brittleness of a weld. Refractory fluxes are sometimes called "fast-freeze" because they solidify at a higher temperature than the weld metal. By becoming solid first, this slag can cradle the molten weld pool and control its shape. This property is very important for out-of-position welds.

The impurities in the weld pool can be metallic or nonmetallic compounds. Metallic elements that are added to the metal during the manufacturing process in small quantities may be concentrated in the weld. These elements improve the grain structure, strength, hardness, resistance to corrosion, or other mechanical properties in the metal's as-rolled or formed state. But the deposited weld metal, or weld nugget, is like a small casting because the liquid weld metal freezes in a controlled shape, and some alloys adversely affect the properties of this casting (weld metal). Nonmetallic compounds are primarily slag inclusions left in the metal from the fluxes used during manufacturing. The welding fluxes form slags that are less dense than the weld metal so that they will float to the surface before the weld solidifies.

Flux Cored Steel Electrode Identification

Module 6
Key Indicator 3

The American Welding Society revised its A5.20 *Specification for Carbon Steel Electrodes for Flux Cored Arc Welding* in 1995 to reflect changes in the composition of the FCA filler metals. Table 3.5 lists the AWS specifications for flux cored filler metals.

Mild Steel

The electrode number *E70T-10* is used as an example to explain the classification system for mild steel FCAW electrodes (Figure 3.14):

- *E*—Electrode.
- *7*—Tensile strength in units of 10,000 psi for a good weld. This value is usually either *6* for 60,000 or *7* for 70,000 psi minimum

Table 3.5 Filler Metal Classification Numbers

Metal	AWS Filler Metal Classification
Mild steel	A5.20
Stainless steel	A5.22
Chromium–molybdenum	A5.29

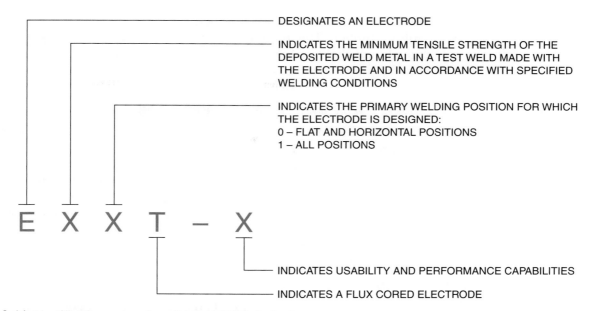

Figure 3.14 Identification system for mild steel FCAW electrodes
Source: Courtesy of the American Welding Society

weld strength. An exception is for the number *12,* which is used to denote filler metals having a range from 70,000 to 90,000 psi.

- *0—0* is used for flat and horizontal fillets only, and *1* is used for all-position electrodes.
- *T*—Tubular (flux cored) electrode.
- *10*—The number in this position can range from *1* to *14* and is used to indicate the electrode's shielding gas if any, number of passes that may be applied one on top of the other, and other welding characteristics of the electrode. The letter *G* is used to indicate that the shielding gas, polarity, and impact properties are not specified. The letter *G* may or may not be followed by the letter *S. S* indicates an electrode suitable only for single-pass welding.

The electrode classification *E70T-10* can have some optional identifiers added to the end of the number, as in *E70T-10MJH8*. These additions are used to add qualifiers to the general classification so that specific codes or standards can be met. These additions have the following meanings:

- *M*—Mixed gas of 75% to 80% Ar and CO_2 for the balance. If there is no *M,* either the shielding gas is CO_2 or the electrode is self-shielded.
- *J*—Describes the Charpy V-notch impact test value of 20 ft-lb at 40°F.
- *H8*—Describes the residual hydrogen levels in the weld: *H4* equals less than 4 ml/100 g; *H8,* less than 8 ml/100 g; *H16,* less than 16 ml/100 g.

Stainless Steel Electrodes

The AWS classification for stainless steel for FCAW electrodes starts with the letter *E* as its prefix. Following the *E* prefix, the American Iron and Steel Institute's (AISI) three-digit stainless steel number is used. This number indicates the type of stainless steel in the filler metal.

To the right of the AISI number, the AWS adds a dash followed by a suffix number. The number 1 is used to indicate an all-position filler metal, and the number 3 is used to indicate an electrode to be used in the flat and horizontal positions only.

Metal Cored Steel Electrode Identification

The addition of metal powders to the flux core of FCA welding electrodes has produced a new classification of filler metals. The new filler metals evolved over time, and a new identification system was established by the AWS to identify these filler metals. Some of the earlier flux cored filler metals that already had powder metals in their core had their numbers changed to reflect the new designation. The designation was changed from the letter *T* for *tubular* to the letter *C* for *core*. For example, E70T-1 became E70C-3C. The complete explanation of the cored electrode *E70C-3C* follows:

- *E*—Electrode.
- *7*—Tensile strength in units of 10,000 psi for a good weld. This value is usually either *6* for 60,000 or *7* for 70,000 psi minimum weld strength. An exception is for the number *12*, which is used to denote filler metals having a range from 70,000 to 90,000 psi.
- *0*—*0* is used for flat and horizontal fillets only, and *1* is used for all-position electrodes.
- *C*—Metal-cored (tubular) electrode.
- *3*—*3* is used for a Charpy impact of 20 ft-lb at 0°F, and *6* represents a Charpy impact of 20 ft-lb at 20°F.
- *C*—The second letter *C* indicates CO_2. The letter *M* in this position would indicate a mixed gas, 75% to 80% Ar, with the balance being CO_2. If there is no *M* or *C*, then the shielding gas is CO_2. The letter *G* is used to indicate that the shielding gas, polarity, and impact properties are not specified. The letter *G* may or may not be followed by the letter *S*. *S* indicates an electrode suitable only for single-pass welding.

Care of Flux Core Electrodes

Wire electrodes may be wrapped in sealed plastic bags for protection from the elements. Others may be wrapped in a special paper, and some are shipped in cans or cardboard boxes.

A small paper bag of a moisture-absorbing material, crystal desiccant, is sometimes placed in the shipping containers to protect wire electrodes from moisture. Some wire electrodes require storage in an electric rod oven to prevent contamination from excessive moisture. Read the manufacturer's recommendations located in or on the electrode shipping container for information on use and storage.

Weather conditions affect your ability to make high-quality welds. Humidity increases the chance of moisture entering the weld zone. Water (H_2O), which consists of two parts hydrogen and one part oxygen, separates in the weld pool. When only one part of hydrogen is expelled, hydrogen entrapment occurs. Hydrogen entrapment can cause weld beads to crack or become brittle. The evaporating moisture will also cause porosity.

NOTE

The powdered metal added to the core flux can provide additional filler metal and/or alloys. This is one way the micro-alloys can be added in very small and controlled amounts, as low as 0.0005% to 0.005%. These are very powerful alloys that dramatically improve the metal's mechanical properties.

To prevent hydrogen entrapment, porosity, and atmospheric contamination, it may be necessary to preheat the base metal to drive out moisture. Storing the wire electrode in a dry location is recommended. The electrode may develop restrictions due to the tangling of the wire or become oxidized with excessive rusting if the wire electrode package is mishandled, thrown, dropped, or stored in a damp location.

SHIELDING GAS

FCA welding wire can be manufactured so that all of the required shielding of the molten weld pool is provided by the vaporization of some of the flux within the tubular electrode. When the electrode provides all of the shielding, it is called **self-shielding** and the welding process is abbreviated FCAW-S (S for self-shielding). Other FCA welding wire must use an externally supplied shielding gas to provide the needed protection of the molten weld pool. When a shielding gas is added, the combined shielding is called **dual shield** and the process is abbreviated FCAW-G (G for gas).

Note: Sometimes the shielding gas(es) are referred to as the shielding *medium*. For example, the shielding gas, or medium, for E71T-5 is either 75% argon with 25% CO_2 or 100% CO_2.

Care must be taken to use the cored electrodes with the recommended gases, and not to use gas at all with the self-shielded electrodes. Using a self-shielding flux cored electrode with a shielding gas may produce a defective weld. The shielding gas will prevent the proper disintegration of much of the deoxidizers. This results in the transfer of these materials across the arc to the weld. In high concentrations, the deoxidizers can produce slags that become trapped in the welds, causing undesirable defects. Lower concentrations may cause brittleness only. In either case, the chance of weld failure is increased. If these electrodes are used correctly, there is no problem.

The selection of a shielding gas will affect the arc and weld properties. The weld bead width, buildup, penetration, spatter, chemical composition, and mechanical properties are all affected as a result of the shielding gas selection.

Shielding gas comes in high-pressure cylinders. These cylinders are supplied with 2000 psi of pressure. Because of this high pressure, it is important that the cylinders be handled and stored safely. For specific cylinder safety instructions see Chapter 2 in Introduction to Welding, the first book in this series.

Gases used for FCA welding include CO_2 and mixtures of argon and CO_2. Argon gas is easily ionized by the arc. Ionization results in a highly concentrated path from the electrode to the weld. This concentration results in a smaller droplet size that is associated with the axial spray mode of metal transfer, Figure 3.15. A smooth, stable arc results and there is a minimum of spatter. This transfer mode continues as CO_2 is added to the argon until the mixture contains more than 25% of CO_2.

As the percentage of CO_2 increases in the argon mixture, weld penetration increases. This increase in penetration continues until a 100% CO_2 shielding gas is reached. But as the percentage of CO_2 is increased the arc stability decreases. The less stable arc causes an increase in spatter.

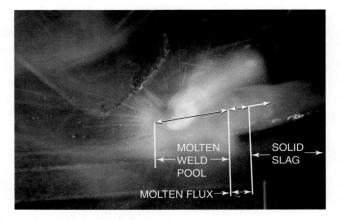

Figure 3.15 Axial spray transfer mode
Source: Courtesy of Larry Jeffus

A mixture of 75% argon and 25% CO_2 works best for jobs requiring a mixed gas. This mixture is sometimes called C-25.

Straight CO_2 is used for some welding. But the CO_2 gas molecule is easily broken down in the welding arc. It forms carbon monoxide (CO) and free oxygen (O). Both gases are reactive to some alloys in the electrode. As these alloys travel from the electrode to the molten weld pool, some of them form oxides. Silicon and manganese are the primary alloys that become oxidized and lost from the weld metal.

Most FCA welding electrodes are specifically designed to be used with or without shielding gas and for a specific shielding gas or percentage mixture. For example, an electrode designed specifically for use with 100% CO_2 will have higher levels of silicon and manganese to compensate for the losses to oxidization. But if 100% argon or a mixture of argon and CO_2 is used, the weld will have an excessive amount of silicon and manganese. The weld will not have the desired mechanical or metallurgical properties. Although the weld may look satisfactory, it will probably fail prematurely.

Caution

Never use an FCA welding electrode with a shielding gas it is not designated to be used with. The weld it produces may be unsafe.

WELDING TECHNIQUES

A welder can control weld beads made by FCA welding by making changes in the techniques used. The following explains how changing specific welding techniques will affect the weld produced.

Gun Angle

Gun angle, work angle, and *travel angle* are terms used to refer to the relation of the gun to the work surface, Figure 3.16. The gun angle can be used to control the weld pool. The electric arc produces an electrical force known as the arc force. The arc force can be used to counteract the gravitational pull that tends to make the liquid weld pool sag or run ahead of the arc. By manipulating the electrode travel angle for the flat and horizontal position of welding to a 20° to 45° angle from the vertical, the weld pool can be controlled. A 40° to 50° angle from the vertical plate is recommended for fillet welds.

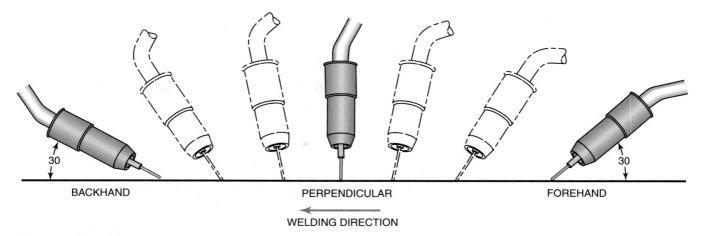

BACKHAND   PERPENDICULAR   FOREHAND

WELDING DIRECTION

Figure 3.16 Welding gun angles

Changes in this angle will affect the weld bead shape and penetration. Shallower angles are needed when welding thinner materials to prevent burn-through. Steeper, perpendicular angles are used for thicker materials.

FCAW electrodes have a flux that is mineral based, often called low-hydrogen. These fluxes are refractory and become solid at a high temperature. If too steep a forehand, or pushing, angle is used, slag from the electrode can be pushed ahead of the weld bead and solidify quickly on the cooler plate, Figure 3.17. Because the slag remains solid at higher temperatures than the temperature of the molten weld pool, it can be trapped under the edges of the weld by the molten weld metal. To avoid this problem, most flat and horizontal welds should be performed with a backhand angle.

Vertical up welds require a forehand gun angle. The forehand angle is needed to direct the arc deep into the groove or joint for better control of the weld pool and deeper penetration, Figure 3.18. Slag entrapment associated with most forehand welding is not a problem for vertical welds.

A gun angle around 90° to the metal surface either slightly forehand or backhand works best for overhead welds, Figure 3.19. The slight angle aids with visibility of the weld, and it helps control spatter buildup in the gas nozzle.

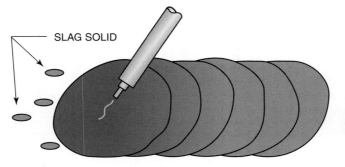

SLAG SOLID

Figure 3.17 Problem of trapped slag
Large quantities of solid slag in front of a weld can cause slag to be trapped under the weld bead.

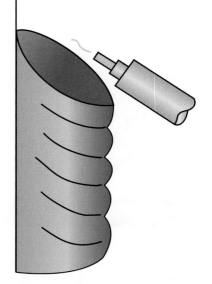

Figure 3.18 Vertical up gun angle

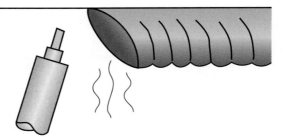

Figure 3.19 Weld gun position to control spatter buildup on an overhead weld

Forehand/Perpendicular/Backhand Techniques

Forehand, perpendicular, and *backhand* are the terms most often used to describe the gun angle as it relates to the work and the direction of travel. The forehand technique is sometimes referred to as *pushing* the weld bead, and backhand may be referred to as *pulling* or *dragging* the weld bead. The term *perpendicular* is used when the gun angle is at approximately 90° to the work surface, Figure 3.20.

Advantages of the Forehand Technique

The forehand welding technique has several advantages:

- Joint visibility—You can easily see the joint where the bead will be deposited, Figure 3.21.
- Electrode extension—The contact tube tip is easier to see, making it easier to maintain a constant extension length.
- Less weld penetration—It is easier to weld on thin sheet metal without melting through.
- Out-of-position welds—This technique works well on vertical up and overhead joints for better control of the weld pool.

Disadvantages of the Forehand Technique

The disadvantages of using the forehand welding technique are the following:

- Weld thickness—Thinner welds may occur because less weld reinforcement is applied to the weld joint.

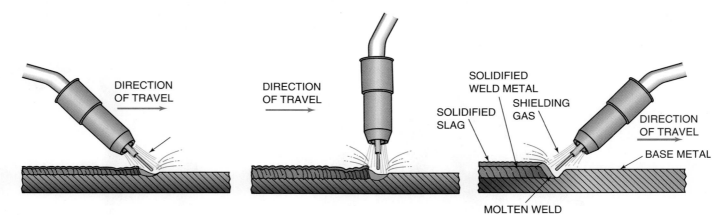

Figure 3.20 Gun angles
Changing the welding gun angle between forehand, perpendicular, and backhand angles will change the shape of the weld bead produced.

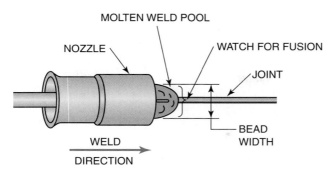

MOLTEN WELD POOL

NOZZLE

WATCH FOR FUSION

JOINT

BEAD
WIDTH

WELD
DIRECTION

Figure 3.21 Welder's view with a forehand angle
This angle keeps the shielding gas nozzle from restricting the welder's view.

- Welding speed—Because less weld metal is being applied, the rate of travel along the joint can be faster, which may make it harder to create a uniform weld.
- Slag inclusions—Some spattered slag can be thrown in front of the weld bead and be trapped or included in the weld, resulting in a weld defect.
- Spatter—Depending on the electrode, the amount of spatter may be slightly increased with the forehand technique.

Advantages of the Perpendicular Technique

The perpendicular welding technique has the following advantages:

- Machine and robotic welding—The perpendicular gun angle is used on automated welding because there is no need to change the gun angle when the weld changes direction.
- Uniform bead shape—The weld's penetration and reinforcement are balanced between those of forehand and backhand techniques.

Disadvantages of the Perpendicular Technique

The disadvantages of using the perpendicular welding technique are the following:

- Limited visibility—Because the welding gun is directly over the weld, there is limited visibility of the weld unless you lean your head way over to the side.
- Weld spatter—Because the weld nozzle is directly under the weld in the overhead position, more weld spatter can collect in the nozzle, causing gas flow problems or even shorting the tip to the nozzle.

Advantages of the Backhand Technique

The backhand welding technique has the following advantages:

- Weld bead visibility—It is easy to see the back of the molten weld pool as you are welding, which makes it easier to control the bead shape, Figure 3.22.
- Travel speed—Because of the larger amount of weld metal being applied, the rate of travel may be slower, making it easier to create a uniform weld.
- Depth of fusion—The arc force and the greater heat from the slower travel rate both increase the depth of weld joint penetration.

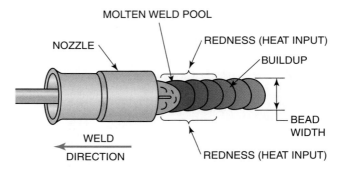

Figure 3.22 Welder's view with a backhand angle
Watch the trailing edge of the molten weld pool.

Disadvantages of the Backhand Technique

The disadvantages of the backhand welding technique are the following:

- Weld buildup—The weld bead may have a convex (raised or rounded) weld face when you use the backhand technique.
- Postweld finishing—Because of the weld bead shape, more work may be required if the product has to be finished by grinding smooth.
- Joint following—It is harder to follow the joint because your hand and the FCAW gun are positioned over the joint, and you may wander from the seam.
- Loss of penetration—An inexperienced welder sometimes directs the wire too far back into the weld pool causing the wire to build up in the face of the weld pool reducing joint penetration.

Travel Speed

The American Welding Society defines travel speed as the linear rate at which the arc is moved along the weld joint. Fast travel speeds deposit less filler metal. If the rate of travel increases, the filler metal cannot be deposited fast enough to adequately fill the path melted by the arc. This causes the weld bead to have a groove melted into the base metal next to the weld and left unfilled by the weld. This condition is known as undercut.

Undercut occurs along the edges or toes of the weld bead. Slower travel speeds will, at first, increase penetration and increase the filler weld metal deposited. As the filler metal increases, the weld bead will build up in the weld pool. Because of the deep penetration of flux cored wire, the angle at which you hold the gun is very important for a successful weld.

If all welding conditions are correct and remain constant, the preferred rate of travel for maximum weld penetration is a travel speed that allows you to stay within the selected welding variables and still control the fluidity of the weld pool. This is an intermediate travel speed, or progression, which is not too fast or too slow.

Another way to figure out correct travel speed is to consult the manufacturer's recommendations chart for the inches per minute (ipm) burn-off rate for the selected electrode.

Mode of Metal Transfer

The mode of metal transfer is used to describe how the molten weld metal is transferred across the arc to the base metal. The mode of metal transfer that is selected, the shape of the completed weld bead, and the depth of weld penetration depend on the welding power source, wire electrode size, type and thickness of material, type of shielding gas used, and best welding position for the task.

Spray Transfer with FCAW-G

The spray transfer mode is the most common process used with gas-shielded FCAW (FCAW-G), Figure 3.15.

As the gun trigger is depressed, the shielding gas automatically flows and the electrode bridges the distance from the contact tube to the base metal, making contact with the base metal to complete a circuit. The electrode shorts and becomes so hot that the base metal melts and forms a weld pool. The electrode melts into the weld pool and burns back toward the contact tube. A combination of high amperage and the shielding gas along with the electrode size produces a pinching effect on the molten electrode wire, causing the end of the electrode wire to spray across the arc.

The characteristic of spray-type transfer is a smooth arc, through which hundreds of small droplets per second are transferred through the arc from the electrode to the weld pool. At that moment a transfer of metal is taking place. Spray transfer can produce a high quantity of metal droplets, up to approximately 250 per second above the transition current, or critical current. This means the current required for a spray transfer to take place is dependent on the electrode size, composition of the electrode, and shielding gas. Below the transition current (critical current), globular transfer takes place.

In order to achieve a spray transfer, high current and larger-diameter electrode wire are needed. A shielding gas of carbon dioxide (CO_2), a mixture of carbon dioxide (CO_2) and argon (Ar), or an argon (Ar) and oxygen (O_2) mixture is needed. FCAW-G is a welding process that, with the correct variables, can be used

- on thin or properly prepared thick sections of material
- on a combination of thick to thin materials
- with small or large electrode diameters
- with a combination of shielding gases

Globular Transfer with FCAW-G

Globular transfer occurs when the welding current is below the transition current, Figure 3.23. The electrode forms a molten ball at its end that grows in size to approximately two to three times the original electrode diameter. These large molten balls are then transferred across the arc at the rate of several drops per second.

The arc becomes unstable because of the gravitational pull from the weight of these large drops. A spinning effect caused by a natural phenomenon takes place when argon gas is introduced to a large ball of molten metal on the electrode. The molten ball spins as it transfers across the arc to the base metal. This unstable globular transfer can produce excessive spatter.

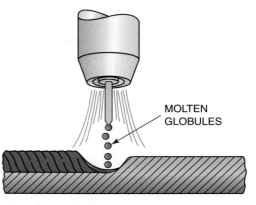

MOLTEN
GLOBULES

Figure 3.23 Globular transfer method

Both FCAW-S and FCAW-G use direct current electrode negative (DCEN) when welding on thin-gauge materials to keep the heat in the base metal and the small-diameter electrode at a controllable burn-off rate. The electrode can then be stabilized, and it is easier to manipulate and control the weld pool in all weld positions. Larger-diameter electrodes are welded with direct current electrode positive (DCEP) because the larger diameters can keep up with the burn-off rates.

The recommended weld position means the position in which the workpiece is placed for welding. All welding positions use either spray or globular transfer, but for now we will concentrate on the flat and horizontal welding positions.

In the flat welding position the workpiece is placed flat on the work surface. In the horizontal welding position the workpiece is positioned perpendicular to the workbench surface.

The amperage range may be from 30 to 400 amperes or more for welding materials from gauge thickness up to 1-1/2 inches. On square groove weld joints, thicker base metals can be welded with little or no edge preparation. This is one of the great advantages of FCAW. If edges are prepared and cut at an angle (beveled) to accept a complete joint weld penetration, the depth of penetration will be greatly increased. FCAW is commonly used for general repairs to mild steel in the horizontal, vertical, and overhead welding positions, sometimes referred to as out-of-position welding.

Electrode Extension

The electrode extension is measured from the end of the electrode contact tube to the point the arc begins at the end of the electrode, Figure 3.24. Compared to GMA welding, the electrode extension required for FCAW is much greater. The longer extension is required for several reasons. The electrical resistance of the wire causes the wire to heat up, which can drive out moisture from the flux. This preheating of the wire also results in a smoother arc with less spatter.

Porosity

FCA welding can produce high-quality welds in all positions, although porosity in the weld can be a persistent problem. Porosity can be caused by moisture in the flux, improper gun manipulation, or surface contamination.

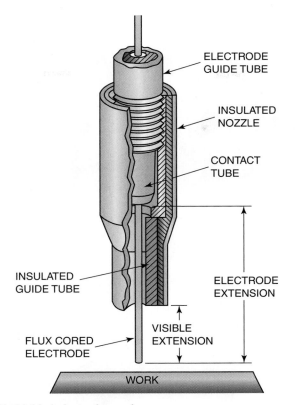

ELECTRODE
GUIDE TUBE

INSULATED
NOZZLE

CONTACT
TUBE

ELECTRODE
EXTENSION

INSULATED
GUIDE TUBE

VISIBLE
EXTENSION

FLUX CORED
ELECTRODE

WORK

Figure 3.24 Self-shielded electrode nozzle
Source: Courtesy of the American Welding Society

The flux used in the FCA welding electrode is subject to picking up moisture from the surrounding atmosphere, so the electrodes must be stored in a dry area. Once the flux becomes contaminated with moisture, it is very difficult to remove. Water (H_2O) breaks down into free hydrogen and oxygen in the presence of an arc, Figure 3.25. The hydrogen can be absorbed into the molten weld metal, where it can cause postweld cracking. The oxygen is absorbed into the weld metal also, but it forms oxides in the metal.

If a shielding gas is used, the FCA welding gun gas nozzle must be close enough to the weld to provide adequate shielding gas coverage. If there is a wind or if the nozzle-to-work distance is excessive, the shielding will be inadequate and allow weld porosity. If welding is to be done outside or in an area subject to drafts, the gas flow rate must be increased or a wind shield must be placed to protect the weld, Figure 3.26.

A common misconception is that the flux within the electrode will either remove or control weld quality problems caused by surface contaminations. That is not true. The addition of flux makes FCA welding more tolerant to surface conditions than GMA welding, although it still is adversely affected by such contaminations.

New hot-rolled steel has a layer of dark gray or black iron oxide called mill scale. Although this layer is very thin, it may provide a source of enough oxygen to cause porosity in the weld. If mill scale causes porosity, it is usually uniformly scattered through the weld, Figure 3.27. Unless it is severe, uniformly scattered porosity is usually not visible in the finished weld. It is trapped under the surface as the weld cools.

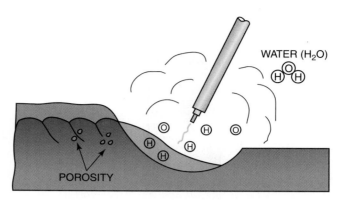

Figure 3.25 Water and porosity
Water (H_2O) breaks down in the presence of the arc and the hydrogen (H) is dissolved in the molten weld metal.

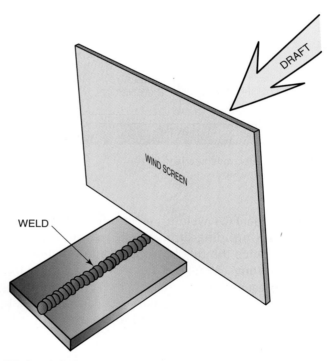

Figure 3.26 Wind and draft protection
A wind screen can keep the welding shielding from being blown away.

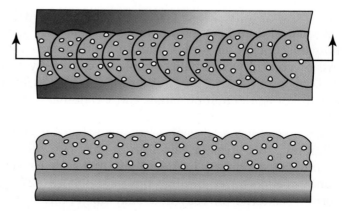

Figure 3.27 Uniformly scattered porosity

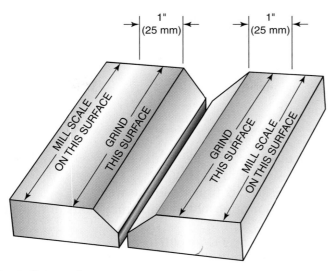

Figure 3.28 Grinding requirements
Grind mill scale off plates within 1 in. (25 mm) of the groove.

Because porosity is under the weld surface, nondestructive testing methods, including X ray, magnetic particle, and ultrasound, must be used to locate it in a weld. It can be detected by mechanical testing such as guided bend, free bend, and nick-break testing for establishing weld parameters. Often it is better to remove the mill scale before welding rather than risking the production of porosity.

All welding surfaces within the weld groove and the surrounding surfaces within 1 in. (25 mm) must be cleaned to bright metal, Figure 3.28. Cleaning may be either grinding, filing, sanding, or blasting.

Any time FCA welds are to be made on metals that are dirty, oily, rusty, or wet or that have been painted, the surface must be precleaned. Cleaning can be done chemically or mechanically.

One advantage of chemically cleaning oil and paint is that it is easier to clean larger areas. Both oil and paint smoke easily when heated, and such smoke can cause weld defects. They must be removed far enough from the weld so that weld heat does not cause them to smoke. In the case of small parts the entire part may need to be cleaned.

SUMMARY

Flux cored arc welding is used to produce more tons of welded fabrications than any other process. The ability to produce high-quality welds on a wide variety of material thicknesses and joint configurations has led to its popularity. As you learn and develop these skills, you will therefore be significantly increasing your employability and productivity in the welding industry.

A wide variety of filler metals and shielding gas combinations for flux cored arc welding are available to you in industry. These various materials aid in producing welds of high quality under various welding conditions. Although the selection of the proper filler metal and gas coverage, if used, will significantly affect the finished weld's quality in the field, there are very few differences in manipulation and setup among these filler metals. Therefore, as you practice welding in a school or training program and learn to use a specific wire and shielding gas mixture, these skills are easily transferable to the next group of materials you will encounter on the job.

REVIEW

1. List some factors that have led to the increased use of FCA welding.
2. How is FCAW similar to GMAW?
3. What does the FCA flux provide to the weld?
4. What are the major atmospheric contaminations of the molten weld metal?
5. How does slag help an FCA weld?
6. What is the electrical difference between a constant-potential and a constant-current power supply?
7. How can FCA welding guns be cooled?
8. What problems does excessive drive roller pressure cause?
9. List the advantages that FCA welding offers the welding industry.
10. Describe the two methods of manufacturing FCA electrode wire.
11. Why are large-diameter electrodes not used for all-position welding?
12. How do deoxidizers remove oxygen from the weld zone?
13. What do fluxing agents do for a weld?
14. Why are alloying elements added to the flux?
15. How does the flux form a shielding gas to protect the weld?
16. What are the main limitations of the rutile fluxes?
17. Why is it more difficult to use lime-based fluxed electrodes on out-of-position welds?
18. What benefit does adding an externally supplied shielding gas have on some rutile-based electrodes?
19. How do excessive amounts of manganese affect a weld?
20. Why are elements added that cause ferrite to form in the weld?
21. Why are some slags called refractory?
22. Why must a flux form a less dense slag?
23. Referring to Table 3.5, what is the AWS classification for FCA welding electrodes for stainless steel?
24. Describe the meaning of each part of the following FCA welding electrode identification: E81T-5.
25. What does the number 316 in E316T-1 mean?
26. What is the advantage of using an argon-CO_2 mixed shielding gas?
27. What are the primary alloying elements lost if 100% CO_2 shielding gas is used?
28. What can cause porosity in an FCA weld?
29. What happens to water in the welding arc?
30. What is the thin, dark gray or black layer on new hot-rolled steel? How can it affect the weld?
31. Why is uniformly scattered porosity hard to detect in a weld?
32. What cautions must be taken when chemically cleaning oil or paint from a piece of metal?
33. What can happen to slag that solidifies on the plate ahead of the weld?
34. How is the electrode extension measured?

Flux Cored Arc Welding

OBJECTIVES

After completing this chapter, the student should be able to

■ set up the FCA weld station

■ thread the electrode wire through the system

■ list three disadvantages of having to bevel a plate before welding

■ make root, filler, and cover passes with the FCAW process

■ make butt welds in all positions that can pass a specified standard's visual or destructive examination criteria

■ make fillet welds in tee joints and lap joints in all positions that can pass a specified standard's visual or destructive examination criteria

KEY TERMS

amperage range	feed rollers	tee joint
conduit liner	lap joint	voltage range
contact tube	root face	weave bead
critical weld	stringer bead	wire-feed speed

AWS SENSE EG2.0

Key Indicators Addressed in this Chapter:

Module 1: Occupational Orientation

Key Indicator 1: Prepares time or job cards, reports, or records

Key Indicator 2: Performs housekeeping duties

Key Indicator 3: Follows verbal instructions to complete work assignments

Key Indicator 4: Follows written instructions to complete work assignments

Module 6: Flux Cored Arc Welding (FCAW-G/GM, FCAW-S)

Key Indicator 1: Performs safety inspections of FCAW equipment and accessories

Key Indicator 2: Makes minor external repairs to FCAW equipment and accessories

Gas Shielded

Key Indicator 3: Sets up for FCAW-G/GM operations on carbon steel

Key Indicator 4: Operates FCAW-G/GM equipment on carbon steel

Key Indicator 5: Makes FCAW-G/GM fillet welds, in all positions, on carbon steel

Key Indicator 6: Makes FCAW-G/GM groove welds, in all positions, on carbon steel

Key Indicator 7: Passes FCAW-G/GM welder performance qualification testing (workmanship sample) on carbon steel

Self Shielded

Key Indicator 8: Sets up for FCAW-S operations on carbon steel

Key Indicator 9: Operates FCAW-S equipment on carbon steel

Key Indicator 10: Makes FCAW-S fillet welds, in all positions, on carbon steel

Key Indicator 11: Makes FCAW-S groove welds, in all positions, on carbon steel

Key Indicator 12: Passes FCAW-S welder performance qualification test (workmanship sample) on carbon steel

Module 9: Welding Inspection and Testing Principles

Key Indicator 1: Examines cut surfaces and edges of prepared base metal parts

Key Indicator 2: Examines tacks, root passes, intermediate layers, and completed welds

INTRODUCTION

Setup of the flux cored arc welding (FCAW) work station is the key to making quality welds. It may be possible, using a poorly set up FCA welder, to make an acceptable weld in the flat position. The FCA welding process is often forgiving; thus welds can often be made even when the welder is not set correctly. However, such welds will have major defects such as excessive spatter, undercut, overlap, porosity, slag inclusions, and poor weld bead contours. Setup becomes even more important for out-of-position welds. Making vertical and overhead welds can be difficult for a student welder with a properly set up system, but it becomes impossible with a system that is out of adjustment.

Learning to set up and properly adjust the FCA welding system will allow you to produce high-quality welds at a high level of productivity.

FCAW is set up and manipulated in a manner similar to that of GMAW. The results of changes in electrode extension, voltage, amperage, and torch angle are essentially the same.

Although every manufacturer's FCA welding equipment is designed differently, all equipment is set up in a similar manner. It is always best to follow the specific manufacturer's recommendations regarding setup as provided in its equipment literature. You will find, however, that, in the field, manufacturers' literature is not always available for the equipment you are asked to use. It is therefore important to have a good general knowledge and understanding of the setup procedure for FCA welding equipment. Figure 4.1 shows all of the various components that make up an FCA welding station.

Caution

FCA welding produces a lot of ultraviolet light, heat, sparks, slag, and welding fumes. Proper personal, protective clothing and special protective clothing must be worn to prevent burns from the ultraviolet light and hot weld metal. Eye protection must be worn to prevent injury from flying sparks and slag. Forced ventilation and possibly a respirator must be used to prevent fume-related injuries. Refer to the safety precautions provided by the equipment and electrode manufacturers and to Chapter 2 in *Welding Skills, Processes and Practices for Entry-Level Welders: Book One* for additional safety help.

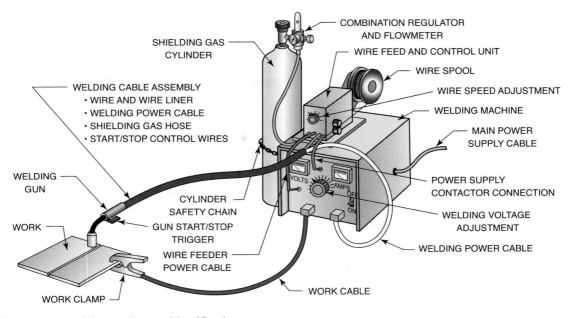

COMBINATION REGULATOR
AND FLOWMETER

SHIELDING GAS
CYLINDER

WIRE FEED AND CONTROL UNIT

WIRE SPOOL

WELDING CABLE ASSEMBLY
• WIRE AND WIRE LINER
• WELDING POWER CABLE
• SHIELDING GAS HOSE
• START/STOP CONTROL WIRES

WIRE SPEED ADJUSTMENT

WELDING MACHINE

MAIN POWER
SUPPLY CABLE

WELDING
GUN

POWER SUPPLY
CONTACTOR CONNECTION

CYLINDER
SAFETY CHAIN

WELDING VOLTAGE
ADJUSTMENT

WORK

GUN START/STOP
TRIGGER

WIRE FEEDER
POWER CABLE

WELDING POWER CABLE

WORK CABLE

WORK CLAMP

VOLTS   AMPS

OFF / ON

Figure 4.1 Basic FCA welding equipment identification

PRACTICES

The practices in this chapter are grouped according to those requiring similar techniques and setups. Plate welds are covered first, then sheet metal. The practices start with 1/4-in. (6-mm) mild steel plates; they are used because they require the least preparation times. The thicker 3/8-in. (9.5-mm) plates provide the basics of practicing groove welding. The 3/4-in. (19-mm) and thicker plates are used to develop the skills required to pass the unlimited thickness test often given to FCA welders. Sheet metal is grouped together because it presents a unique set of learning skills.

The major skill required for making consistently acceptable FCA welds is the ability to set up the welding system. Changes such as variations in material thickness, position, and type of joint require changes both in technique and setup. A correctly set up FCA welding station can, in many cases, be operated by a less-skilled welder. Often the only difference between a welder earning a minimum wage and one earning the maximum wage is the ability to correct machine setups.

For several reasons the FCA welding practice plates will be larger than most other practice plates. Welding heat and welding speed are the major factors that necessitate this increased size. FCA welding is both high energy and fast, and the welding energy (heat) input is so great that small practice plates may glow red by the end of a single weld pass. This would seriously affect the weld quality. To prevent this from happening, wider plates are used. Because of the higher welding speeds, longer plates are usually used.

Plates less than 1/2 in. (13 mm) will be 12 in. (305 mm) long for most practices. In addition to controlling the heat buildup, the longer plates are needed to give the welder enough time to practice welding. Learning to make longer welds is a skill that must also be practiced, because the FCA welding process is used in industry to make long production welds.

Plates thicker than 1/2 in. (13 mm) can be shorter than 12 in. (305 mm). Most codes allow test plates of "unlimited thickness" to be as short as 7 in. (178 mm).

PRACTICE 4-1

FCAW Equipment Setup

For this practice, you will need a semiautomatic welding power source approved for FCA welding, welding gun, electrode feed unit, electrode supply, shielding gas supply, shielding gas flowmeter regulator, electrode conduit, power and work leads, shielding gas hoses (if required), assorted hand tools, spare parts, and any other required materials. In this practice, you will demonstrate to a group of students and your instructor how to properly set up an FCA welding station. Some manufacturers include detailed setup instructions with their equipment. If such instructions are available for your equipment, follow them. Otherwise, use the following instructions.

If the shielding gas is to be used and it comes from a cylinder, the cylinder must be chained securely in place before the valve protection cap is removed, Figure 4.2. Standing to one side of the cylinder, make sure nobody is in line with the valve and quickly crack the valve to blow out any dirt in the valve before the flowmeter regulator is attached, Figure 4.3. Attach the correct hose from the regulator to the "gas-in" connection on the electrode feed unit or machine.

Install the reel of electrode (welding wire) on the holder and secure it, Figure 4.4. Check the feed roller size to ensure that it matches the wire size, Figure 4.5. Also check the **conduit liner** size for compatibility with the wire size. Connect the conduit to the feed unit. The conduit or an extension should be aligned with the groove in the roller and set as close to the roller as possible without touching, Figure 4.6. Misalignment at this point can contribute to a bird's nest, Figure 4.7. Bird-nesting of the electrode wire, so called because it looks like a bird's nest, results when the feed roller pushes the wire into a tangled ball because the wire would not go through the outfeed side conduit.

Figure 4.2 Make sure the gas cylinder is chained securely in place before removing the safety cap
Source: Courtesy of Larry Jeffus

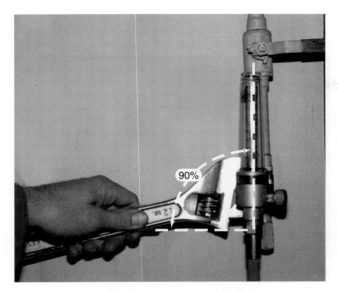

Figure 4.3 Attaching the flowmeter regulator
Be sure the tube is vertical.
Source: Courtesy of Larry Jeffus

Figure 4.4 Wire reel may be secured by a center nut or locking lever
Source: Courtesy of Larry Jeffus

Figure 4.5 Checking feed roller size
Check to be certain that the feed rollers are the correct size for the wire being used.
Source: Courtesy of Larry Jeffus

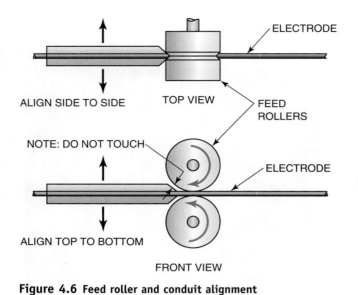

Figure 4.6 Feed roller and conduit alignment

Figure 4.7 "Bird's nest" in the filler wire at the feed rollers
Source: Courtesy of Larry Jeffus

Be sure the power is off before attaching the welding cables. The electrode and work leads should be attached to the proper terminals. The electrode lead should be attached to the electrode or positive (+) terminal. If necessary, it is also attached to the power cable part of the gun lead. The work lead should be attached to the work or negative (–) terminal.

The shielding "gas-out" side of the solenoid is then also attached to the gun lead. If a separate splice is required from the gun switch circuit to the feed unit, it should be connected at this time. Check that the welding contactor circuit is connected from the feed unit to the power source.

The welding cable liner or wire conduit must be securely attached to the gas diffuser and contact tube, Figure 4.8. The **contact tube** (tip), Figure 3.1, must be the correct size to match the electrode wire size being used. If a shielding gas is to be used, a gas nozzle would be attached to complete the assembly. If a gas nozzle is not needed for a shielding gas, it may still be installed. Because it is easy for a student to touch the work with the contact tube during welding, an electrical short may occur. This short-out of the contact tube will immediately destroy the tube. Although the gas nozzle may interfere with some visibility, it may be worth the trouble for a new welder. FCA welding is more sensitive to changes in arc voltage than is SMA (stick) welding. Such variations in FCA welding voltage can dramatically and adversely affect your ability to maintain weld bead control.

A loose or poor connection will result in increased circuit resistance and a loss of welding voltage. To be sure that you have a good work

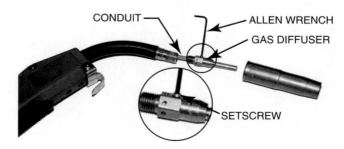

Figure 4.8 Securely attach conduit to gas diffuser and contact tube to prevent wire jams caused by misalignment
Source: Courtesy of Larry Jeffus

connection, remove any dirt, rust, oil, or other surface contamination at the point where the work clamp is connected to the weldment.

Complete a copy of the "Student Welding Report" listed in Appendix I or provided by your instructor.

PRACTICE 4-2

Threading FCAW Wire

Using the FCAW machine that was properly assembled in Practice 4-1, you will turn the machine on and thread the electrode wire through the system.

Check that the unit is assembled correctly according to the manufacturer's specifications. Switch on the power and check the gun switch circuit by depressing the switch. The power source relays, feed relays, gas solenoid, and feed motor should all activate.

Cut off the end of the electrode wire if it is bent. When working with the wire, be sure to hold it tightly. The wire will become tangled if it is released. The wire has a natural curl known as *cast*. Straighten out about 12 in. (300 mm) of the curl to make threading easier. Separate the wire **feed rollers** and push the wire first through the guides, then between the rollers, and finally into the conduit liner, Figure 4.9. Reset the rollers so there is a slight amount of compression on the wire, Figure 4.10. Set the **wire-feed speed** control to a slow speed. Hold the welding gun so that the electrode conduit and cable are as straight as possible.

Wearing safety glasses and pointing the gun away from the welder's face, press the gun switch or the cold feed switch on the feeder. Pressing the gun switch to start the wire feeder is called triggering the gun. The cold feed switch on the feeder is a safety option built into some equipment that advances the wire without current being sent to the gun. The wire should start feeding into the liner. Watch to make certain that the wire feeds smoothly and release the gun switch as soon as the end comes through the gun.

If the wire stops feeding before it reaches the end of the gun, stop and check the system. If no obvious problem can be found, mark the wire with tape and remove it from the gun. It can then be held next to the system to determine the location of the problem.

Module 1
Key Indicator 1, 2, 3, 4

Module 2
Key Indicator 1, 2, 3, 4, 7

Module 6
Key Indicator 2
Gas Shielded
Key Indicator 3
Self Shielded
Key Indicator 8

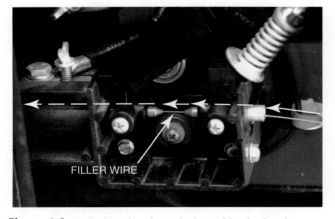

Figure 4.9 Push the wire through the guides by hand
Source: Courtesy of Larry Jeffus

Figure 4.10 Adjust the wire-feed tensioner
Source: Courtesy of Larry Jeffus

With the wire feed running, adjust the feed roller compression so that the wire reel can be stopped easily by a slight pressure. Too light a roller pressure will cause the wire to feed erratically. Too high a pressure can crush some wires, causing some flux to be dropped inside the wire liner. If this happens, you will have a continual problem with the wire not feeding smoothly or jamming.

With the feed running, adjust the spool drag so that the reel stops when the feed stops. The reel should not coast to a stop, because it allows slack in the wire that can easily be snagged. Also, when the feed restarts, a jolt occurs when the slack in the wire is taken up. This jolt can be enough to momentarily stop the wire, possibly causing a discontinuity in the weld.

When the test runs are completed, you can either rewind or cut off the wire. Some wire-feed units have a retract button. This allows the feed driver to reverse and retract the wire automatically. To rewind the wire on units without this retraction feature, release the rollers and turn them backward by hand. If the machine will not allow the feed rollers to be released without upsetting the tension, you must cut the wire. Some wire reels have covers to prevent the collection of dust, dirt, and metal filings on the wire, Figure 4.11.

Complete a copy of the "Student Welding Report" listed in Appendix I or provided by your instructor.

(A)

(B)

Figure 4.11 Wire protection
(A) Covered wire reel. (B) Wire cover on a dual wire-feed system.
Source: Courtesy of Lincoln Electric Company

FLAT-POSITION WELDS
PRACTICE 4-3

Stringer Beads, Flat Position

Using a properly set up and adjusted FCA welding machine, Table 4.1, proper safety protection, E70T-1 and/or E71T-11 electrodes of diameter 0.035 in. and/or 0.045 in. (0.9 mm and/or 1.2 mm), and one or more pieces of mild steel plate, 12 in. (305 mm) long and 1/4 in. (6 mm) or thicker, you will make a stringer bead weld in the flat position, Figure 4.12.

Module 1
Key Indicator 1, 2, 3, 4

Module 2
Key Indicator 1, 2, 3, 4, 7

Module 6
Key Indicator 1
Gas Shielded
Key Indicator 4
Self Shielded
Key Indicator 9

Table 4.1 FCA Welding Parameters for Use if Specific Settings Are Unavailable from Electrode Manufacturer (base metal thickness 1/4 to 1/2 inch)

Electrode		Welding Power			Shielding Gas		Base Metal	
Type	Size	Amps	Wire-feed Speed, ipm (cm/min.)	Volts	Type	Flow	Type	Thickness
E70T-1 E71T-1	0.035 in. (0.9 mm)	130 to 150	288 to 380 (732 to 975)	22 to 25	None	n/a	Low-carbon steel	1/4 in. to 1/2 in. (6 mm to 13 mm)
E70T-1 E71T-1	0.045 in. (1.2 mm)	150 to 210	200 to 300 (508 to 762)	28 to 29	None	n/a	Low-carbon steel	1/4 in. to 1/2 in. (6 mm to 13 mm)
E70T-5 E71T-11	0.035 in. (0.9 mm)	130 to 200	288 to 576 (732 to 1463)	20 to 28	75% argon 25% CO_2	30 cfh	Low-carbon steel	1/4 in. to 1/2 in. (6 mm to 13 mm)
E70T-5 E71T-11	0.045 in. (1.2 mm)	150 to 250	200 to 400 (508 to 1016)	23 to 29	75% argon 25% CO_2	35 cfh	Low-carbon steel	1/4 in. to 1/2 in. (6 mm to 13 mm)

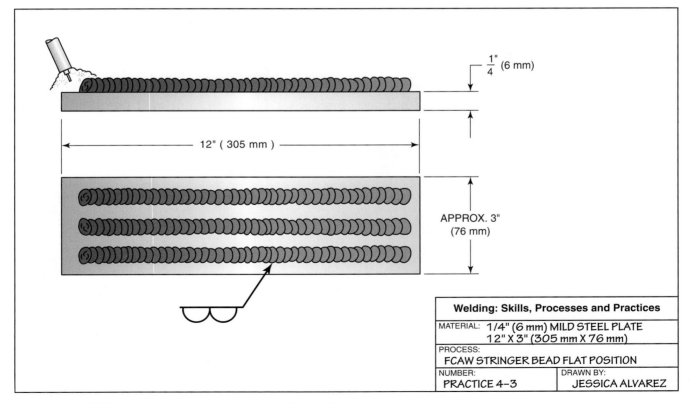

Figure 4.12 FCAW stringer bead, 1/4 in. mild steel, flat position

Starting at one end of the plate and using a dragging technique, make a weld bead along the entire 12-in. (305-mm) length of the metal. After the weld is complete, check its appearance. Make any needed changes to correct the weld. Repeat the weld and make additional adjustments. After the machine is set, start to work on improving the straightness and uniformity of the weld. Use weave patterns of different widths and straight stringers without weaving.

Repeat with both classifications of electrodes until beads can be made straight, uniform, and free from any visual defects. Turn off the welding machine and shielding gas and clean up your work area when you are finished welding.

Complete a copy of the "Student Welding Report" listed in Appendix I or provided by your instructor.

SQUARE-GROOVE WELDS

One advantage of FCA welding is the ability to make 100%-joint-penetrating welds without beveling the edges of the plates. These full-joint-penetrating welds can be made in plates that are 1/4 in. (6 mm) or less in thickness. Welding on thicker plates risks the possibility of a lack of fusion on both sides of the **root face**, Figure 4.13.

There are several disadvantages of having to bevel a plate before welding:

* Beveling the edge of a plate adds an operation to the fabrication process.
* Both more filler metal and welding time are required to fill a beveled joint than are required to make a square jointed weld.
* Beveled joints have more heat from the thermal beveling and additional welding required to fill the groove. The lower heat input to the square joint means less distortion.

The major disadvantage of making square jointed welds is that as the plate thickness approaches 1/4 in. (6 mm) or if the weld is out of

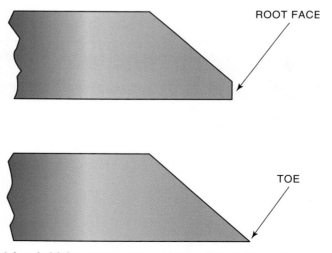

Figure 4.13 A beveled joint may or may not have a flat surface, called a root face

position, a much higher level of skill is required. The skill required to make quality square welds can be acquired by practicing on thinner metal. It is much easier to make this type of weld in metal 1/8 in. (3 mm) thick and then move up in thickness as your skills improve.

PRACTICE 4-4

Butt Joint 1G

Using a properly set up and adjusted FCA welding machine, proper safety protection, E70T-1 and/or E71T-11 electrodes of diameter 0.035 in. and/or 0.045 in. (0.9 mm and/or 1.2 mm), and one or more pieces of mild steel plate, 12 in. (305 mm) long and 1/4 in. (6 mm) or less in thickness, you will make a groove weld in the flat position, Figure 4.14.

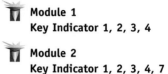

Module 1
Key Indicator 1, 2, 3, 4

Module 2
Key Indicator 1, 2, 3, 4, 7

Module 6
Gas Shielded
Key Indicator 6
Self Shielded
Key Indicator 11
This practice addresses the "Flat" position portion of the all-position requirement of 6 and 11.

- Tack weld the plates together and place them in position to be welded.
- Starting at one end, run a bead along the joint. Watch the molten weld pool and bead for signs that a change in technique may be required.
- Make any needed changes as the weld progresses in order to produce a uniform weld.

Repeat with both classifications of electrodes until defect-free welds can consistently be made in the 1/4-in.-thick (6-mm-thick) plate. Turn

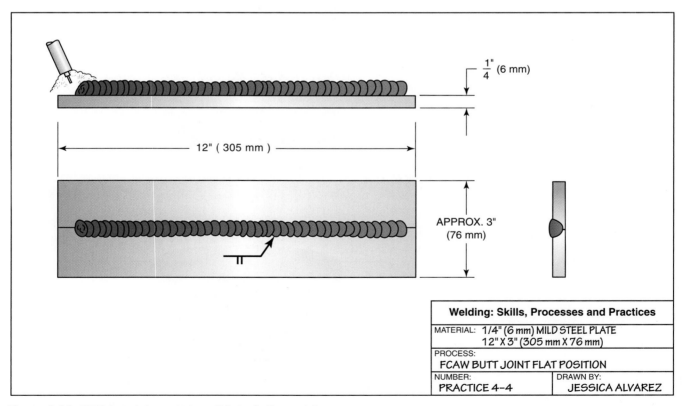

Welding: Skills, Processes and Practices

MATERIAL: 1/4" (6 mm) MILD STEEL PLATE 12" X 3" (305 mm X 76 mm)	
PROCESS: FCAW BUTT JOINT FLAT POSITION	
NUMBER: PRACTICE 4–4	DRAWN BY: JESSICA ALVAREZ

Figure 4.14 FCAW butt joint, 1/4 in. mild steel, flat position

off the welding machine and shielding gas and clean up your work area when you are finished welding.

Complete a copy of the "Student Welding Report" listed in Appendix I or provided by your instructor.

V-GROOVE AND BEVEL-GROOVE WELDS

Although for speed and economy engineers try to avoid specifying welds that require beveling the edges of plates, it is not always possible. Anytime the metal being welded is thicker than 1/4 in. (6 mm) and a 100% joint penetration weld is required, the edges of the plate must be prepared with a bevel. Fortunately, FCA welding allows a narrower groove to be made and still achieve a thorough thickness weld, due to the deeper penetration characteristics of the FCAW process, Figure 4.15.

All FCA groove welds are made using three different types of weld passes, Figure 4.16.

- *Root pass:* The first weld bead of a multiple-pass weld. The root pass fuses the two parts together and establishes the depth of weld metal penetration.
- *Filler pass:* Made after the root pass is completed and used to fill the groove with weld metal. More than one pass is often required.
- *Cover pass:* The last weld pass on a multipass weld. The cover pass may be made with one or more welds. It must be uniform in width, reinforcement, and appearance.

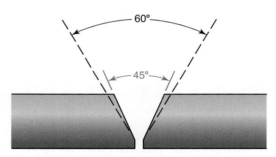

Figure 4.15 Reduced groove angle for FCAW
A smaller groove angle reduces both weld time and filler metal required to make the weld.

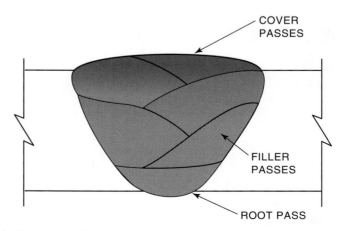

Figure 4.16 Three types of weld passes make up a weld

Root Pass

A good root pass is needed in order to obtain a sound weld. The root may be either open or closed using a backing strip, Figure 4.17.

The backing strips are usually made from a piece of metal 1/4 in. (6 mm) thick and 1 in. (25 mm) wide, and should be 2 in. (50 mm) longer than the base plates. The strip is attached to the plate by tack welds made on the sides of the strip, Figure 4.18.

Most production welds do not use backing strips, so they are made as open root welds. Because of the difficulty in controlling root weld face contours in FCAW, however, open-root joints are often avoided on **critical welds**. If an open-root weld is needed because of weldment design, the root pass may be put in with an SMAW electrode or the root face of the FCA weld can be retouched by grinding and/or back welding.

Care must be taken with any root pass not to have the weld face too convex, Figure 4.19. Convex weld faces tend to trap slag along the toe of

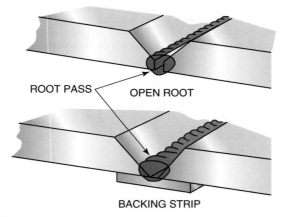

Figure 4.17 Root pass
The maximum deposit for a root pass is 1/4 in. (6 mm) thick.

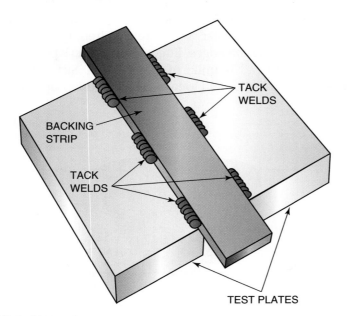

Figure 4.18 Backing strip
Securely tack weld the backing strip to the test plates.

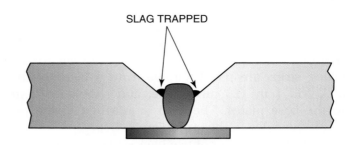

Figure 4.19 Slag trapped beside weld bead is hard to remove

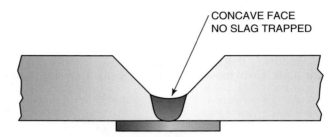

Figure 4.20 Flat or concave weld faces are easier to clean off

the weld. FCA weld slag can be extremely difficult to remove in this area, especially if there is any undercutting. To avoid this, adjust the welding power settings, speed, and weave pattern so that a flat or slightly concave weld face is produced, Figure 4.20.

Filler Pass

Filler passes are made with either **stringer beads**, which are made with a straight progression and very little gun manipulation, or **weave beads**, in which the operator oscillates the gun from side to side in order to widen the weld profile. Either bead type works well for flat or vertically positioned welds, but stringer beads work best for horizontal and overhead-positioned welds. When multiple-pass filler welds are required, each weld bead must overlap the others along the edges. Edges should overlap smoothly enough so that the finished bead is uniform, Figure 4.21. Stringer beads usually overlap about 25% to 50%, and weave beads overlap approximately 10% to 25%.

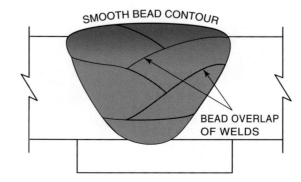

Figure 4.21 The surface of a multipass weld should be as smooth as if it were made by one weld

Each weld bead must be cleaned before the next bead is started. The filler pass ends when the groove has been filled to a level just below the plate surface.

Cover Pass

The cover pass may or may not simply be a continuation of the weld beads used to make the filler pass(es). The major difference between the filler pass and the cover pass is the weld face importance. Keeping the face and toe of the cover pass uniform in width, reinforcement, and appearance and free of defects is essential. Most welds are not tested beyond a visual inspection. For that reason the appearance might be the only factor used for accepting or rejecting welds.

The cover pass must meet a strict visual inspection standard. The visual inspection looks to see that the weld is uniform in width and reinforcement. There should be no arc strikes or hammer marks from chipping or slag removal operations on the plate other than those on the weld itself. The weld must be free of both incomplete fusion and cracks. The weld must be free of overlap, and undercut must not exceed either 10% of the base metal or 1/32 in. (0.8 mm), whichever is less. Reinforcement must have a smooth transition with the base plate and be no higher than 1/8 in. (3 mm), Figure 4.22.

PRACTICE 4-5

Butt Joint 1G

Use a properly set up and adjusted FCA welding machine, Table 4.1; proper safety protection; E70T-1 and/or E71T-11 electrodes of diameter 0.035 in. and/or 0.045 in. (0.9 mm and/or 1.2 mm); one or more pieces of mild steel plate, beveled, 12 in. (305 mm) long and 3/8 in. (9.5 mm) thick; and a backing strip 14 in. (355 mm) long, 1 in. (25 mm) wide, and 1/4 in. (6 mm) thick. You will make a groove weld in the flat position, Figure 4.23.

Tack weld the backing strip to the plates. There should be a root gap of approximately 1/8 in. (3 mm) between the plates. The beveled surface can be made with or without a root face, Figure 4.24.

Module 1
Key Indicator 1, 2, 3, 4

Module 2
Key Indicator 1, 2, 3, 4, 7

Module 6
Gas Shielded
Key Indicator 6
Gas Shielded
Key Indicator 11
This practice addresses the "Flat" position portion of the all-position requirement of 6 and 11.

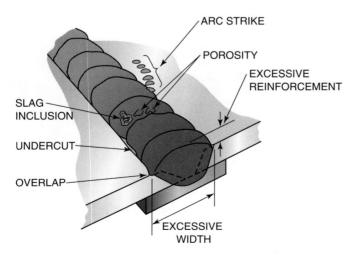

Figure 4.22 Common discontinuities found during a visual examination

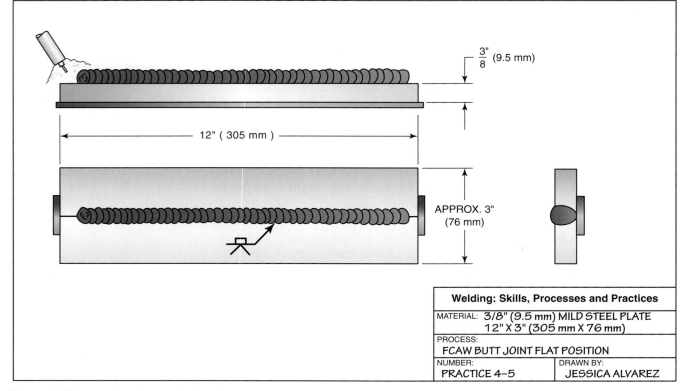

Figure 4.23 FCAW butt joint, 3/8 in. mild steel, flat position

Place the test plates in position at a comfortable height and location. Be sure that you have complete and free movement along the full length of the weld joint. It is often a good idea to make a practice pass along the joint with the welding gun without power to make sure nothing will interfere with your making the weld. Be sure the welding cable is free and will not get caught on anything during the weld.

Start the weld outside the groove on the backing strip tab, Figure 4.25. This is done so that the arc is smooth and the molten weld pool size is

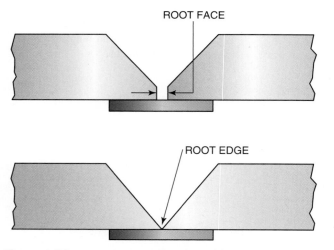

Figure 4.24 Groove layout with and without a root face

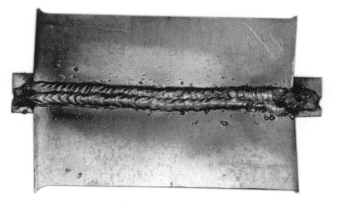

Figure 4.25 Using the ends of a backing strip
Run-off tabs, in which the weld starts and stops on the ends of the backing strip, help control possible underfill or burn-back at the starting and stopping points of a groove weld.
Source: Courtesy of Larry Jeffus

established at the beginning of the groove. Continue the weld out onto the tab at the outer end of the groove. This process ensures that the end of the groove is completely filled with weld.

Repeat with both classifications of electrodes until consistently defect-free welds can be made. Turn off the welding machine and shielding gas and clean up your work area when you are finished welding.

Complete a copy of the "Student Welding Report" listed in Appendix I or provided by your instructor.

PRACTICE 4-6

Butt Joint 1G 100% to Be Tested

Use a properly set up and adjusted FCA welding machine; proper safety protection; E70T-1 and/or E71T-11 electrodes of diameter 0.035 in. and/or 0.045 in. (0.9 mm and/or 1.2 mm); one or more pieces of mild steel plate, beveled, 12 in. (305 mm) long and 3/8 in. (9.5 mm) thick; and a backing strip 14 in. (355 mm) long, 1 in. (25 mm) wide, and 1/4 in. (6 mm) thick. You will make a groove weld in the flat position, Figure 4.26.

Following the same instructions for the assembly and welding procedure outlined in Practice 4-5, repeat the weld until you can use each electrode type to make welds with 100% penetration that will pass a bend test. Turn off the welding machine and shielding gas and clean up your work area when you are finished welding.

Complete a copy of the "Student Welding Report" listed in Appendix I or provided by your instructor.

Module 1
Key Indicator 1, 2, 3, 4

Module 2
Key Indicator 1, 2, 3, 4, 7

Module 6
Gas Shielded
Key Indicator 6
Self Shielded
Key Indicator 11
This practice addresses the "Flat" position portion of the all-position requirement of 6 and 11.

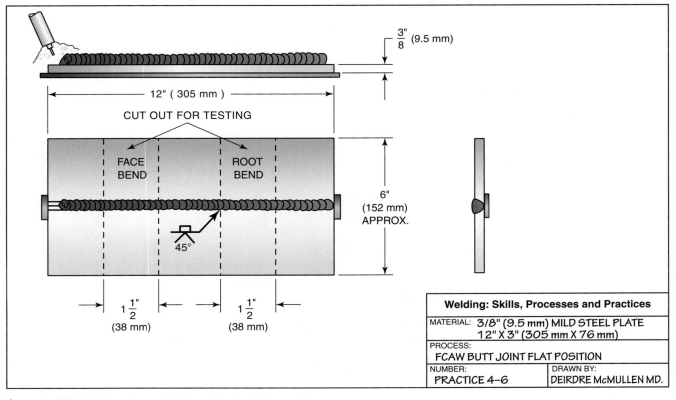

Figure 4.26 FCAW butt joint, 3/8 in. mild steel, flat position

FILLET WELDS

A fillet weld is the type of weld made on a **lap joint** and a **tee joint**. It should be built up equal to the thickness of the plate, Figure 4.27. On thick plates the fillet must be made up of several passes as with a groove weld. The difference with a fillet weld is that a smooth transition from the plate surface to the weld is required. If this transition is abrupt, it can cause stresses that will weaken the joint.

The lap joint is made by overlapping the edges of the plates. They should be held together tightly before tack welding them together. A small tack weld may be added in the center to prevent distortion during welding, Figure 4.28. Chip and wire brush the tacks before you start to weld.

The tee joint is made by tack welding one piece of metal on another piece of metal at a right angle, Figure 4.29. After the joint is tack welded together, the slag is chipped from the tack welds. If the slag is not removed, it will cause a slag inclusion in the final weld.

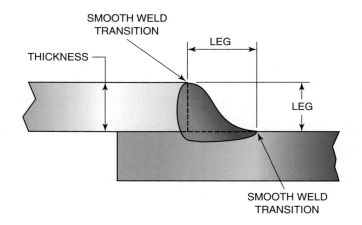

Figure 4.27 The legs of a fillet weld should generally be equal to the thickness of the base metal

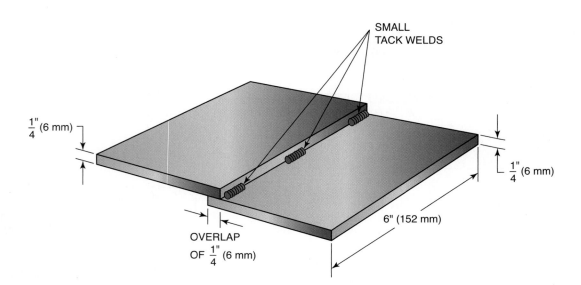

Figure 4.28 Tack welding the plates of a lap joint together

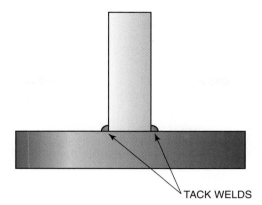

Figure 4.29 Tack welding both sides of a tee joint
Tack welding will help keep the tee square for welding.

Holding thick plates tightly together on tee joints may cause underbead cracking, or lamellar tearing, Figure 4.30. On thick plates the weld shrinkage can be great enough to pull the metal apart well below the bead or its heat-affected zone. In production welds, cracking can be controlled by not assembling the plates tightly together. The space between the two plates can be set by placing a small wire spacer between them, Figure 4.31.

A fillet welded lap or tee joint can be strong if it is welded on both sides, even without having deep penetration, Figure 4.32. Some tee joints may be prepared for welding by cutting either a bevel or a J-groove in the vertical plate. This cut is not required for strength but may be necessary because of design limitations. Unless otherwise specified, most fillet welds will be equal in size to the plates welded. A fillet weld will be as strong as the base plate if the size of the two welds equals the total thickness of the base plate. The weld bead should have a flat or slightly concave appearance to ensure the greatest strength and efficiency, Figure 4.33.

The root of fillet welds must be melted to ensure a completely fused joint. A notch along the root of the weld pool is an indication that the root is not being fused together, Figure 4.34. To achieve complete root fusion, move the arc to a point as close as possible to the leading edge of the weld pool, Figure 4.35. If the arc strikes the unmelted plate ahead of the molten weld pool, it may become erratic, which will increase weld spatter.

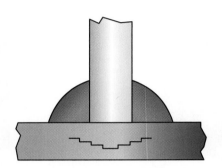

Figure 4.30 Underbead cracking, or lamellar tearing, of the base plate

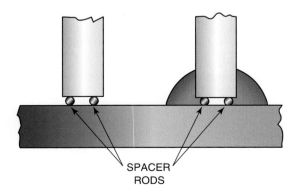

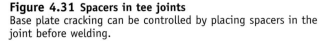

Figure 4.31 Spacers in tee joints
Base plate cracking can be controlled by placing spacers in the joint before welding.

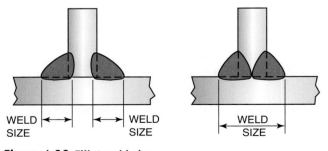

Figure 4.32 Fillet weld size
If the total weld sizes are equal, then both tee joints would have equal strength.

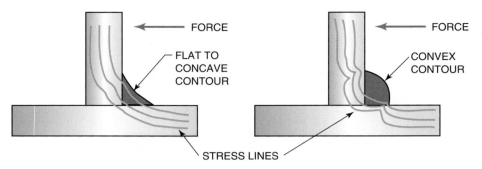

Figure 4.33 Fillet weld shape
The stresses are distributed more uniformly through a flat or concave fillet weld.

Figure 4.34 Watch the root of the weld bead to be sure there is complete fusion
Source: Courtesy of Larry Jeffus

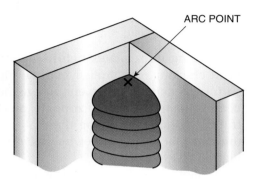

Figure 4.35 Moving the arc as close as possible to the leading edge of the weld will provide good root fusion.

PRACTICE 4-7

Lap Joint and Tee Joint 1F

Module 1
Key Indicator 1, 2, 3, 4

Module 2
Key Indicator 1, 2, 3, 4, 7

Module 6
Gas Shielded
Key Indicator 5
Self Shielded
Key Indicator 10
This practice addresses the "Flat" position portion of the all-position requirement for 5 and 10.

Use a properly set up and adjusted FCA welding machine; proper safety protection; E70T-1 and/or E71T-11 electrodes of diameter 0.035 in. and/or 0.045 in. (0.9 mm and/or 1.2 mm); and one or more pieces of mild steel plate, beveled, 12 in. (305 mm) long and 3/8 in. (9.5 mm) thick. You will make a fillet weld in the flat position.

Tack weld the pieces of metal together and brace them in position. When making the lap or tee joints in the flat position, the plates must be at a 45° angle so that the surface of the weld will be flat, Figure 4.36A and Figure 4.36B. Starting at one end, make a weld along the entire length of the joint.

Repeat each type of joint with both classifications of electrodes until consistently defect-free welds can be made. Turn off the welding machine and shielding gas and clean up your work area when you are finished welding.

Complete a copy of the "Student Welding Report" listed in Appendix I or provided by your instructor.

PRACTICE 4-8

Tee Joint 1F

Module 1
Key Indicator 1, 2, 3, 4

Module 2
Key Indicator 1, 2, 3, 4, 7

Module 6
Gas Shielded
Key Indicator 5
Self Shielded
Key Indicator 10

Use a properly set up and adjusted FCA welding machine, Table 4.2; proper safety protection; E70T-1 and/or E71T-11 electrodes of diameter 0.035 in. and/or through 1/16 in. (0.9 mm and/or through 1.6 mm); one

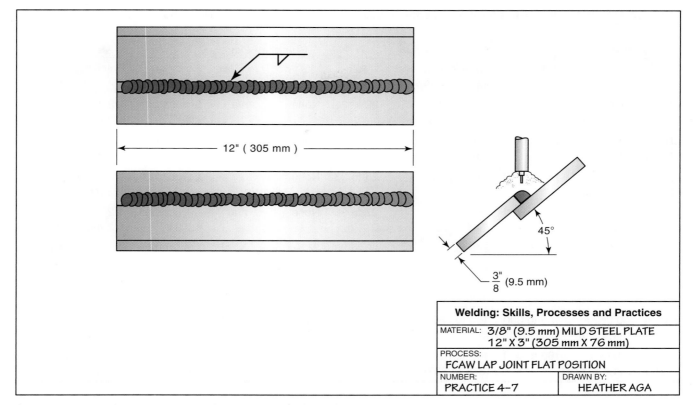

Figure 4.36 (A) FCAW lap joint, 3/8 in. mild steel, flat position

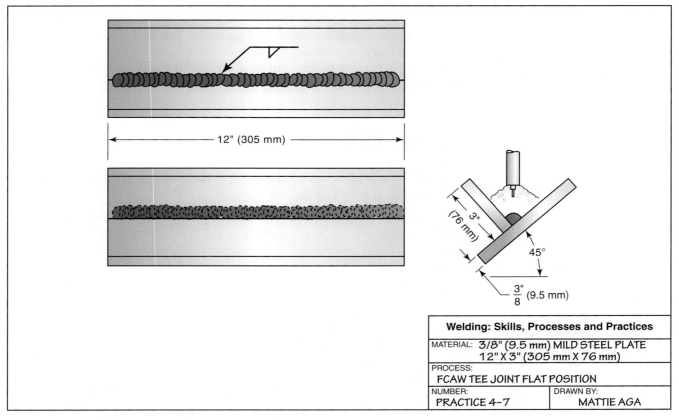

Figure 4.36 (B) FCAW tee joint, 3/8 in. mild steel, flat position

Table 4.2 FCA Welding Parameters for Use if Specific Settings Are Unavailable from Electrode Manufacturer (base metal thickness 1/2 to 3/4 inch)

Electrode		Welding Power			Shielding Gas		Base Metal	
Type	Size	Amps	Wire-feed Speed, ipm (cm/min)	Volts	Type	Flow	Type	Thickness
E70T-1 E71T-1	0.035 in. (0.9 mm)	130 to 150	288 to 380 (732 to 975)	22 to 25	None	n/a	Low-carbon steel	1/2 in. to 3/4 in. (13 mm to 19 mm)
E70T-1 E71T-1	0.045 in. (1.2 mm)	150 to 210	200 to 300 (508 to 762)	28 to 29	None	n/a	Low-carbon steel	1/2 in. to 3/4 in. (13 mm to 19 mm)
E70T-1 E71T-1	.052 in. (1.4 mm)	150 to 300	150 to 350 (381 to 889)	25 to 33	None	n/a	Low-carbon steel	1/2 in. to 3/4 in. (13 mm to 19 mm)
E70T-1 E71T-1	1/16 in. (1.6 mm)	200 to 400	150 to 300 (381 to 762)	27 to 33	None	n/a	Low-carbon steel	1/2 in. to 3/4 in. (13 mm to 19 mm)
E70T-5 E71T-11	0.035 in. (0.9 mm)	130 to 200	288 to 576 (732 to 1463)	20 to 28	75% argon 25% CO_2	30 cfh	Low-carbon steel	1/2 in. to 3/4 in. (13 mm to 19 mm)
E70T-5 E71T-11	0.045 in. (1.2 mm)	150 to 250	200 to 400 (508 to 1016)	23 to 29	75% argon 25% CO_2	35 cfh	Low-carbon steel	1/2 in. to 3/4 in. (13 mm to 19 mm)
E70T-5 E71T-11	0.052 in. (1.4 mm)	150 to 300	150 to 350 (381 to 889)	21 to 32	75% argon 25% CO_2	35 cfh	Low-carbon steel	1/2 in. to 3/4 in. (13 mm to 19 mm)
E70T-5 E71T-11	1/16 in. (1.6 mm)	180 to 400	145 to 350 (368 to 889)	21 to 34	75% argon 25% CO_2	40 cfh	Low-carbon steel	1/2 in. to 3/4 in. (13 mm to 19 mm)

or more pieces of mild steel plate, beveled, 7 in. (178 mm) long and 3/4 in. (19 mm) thick or thicker. You will make a fillet weld in the flat position.

Following the same instructions for the assembly and welding procedure outlined in Practice 4-7, repeat each type of joint with both classifications of electrodes until consistently defect-free welds can be made. Turn off the welding machine and shielding gas and clean up your work area when you are finished welding.

Complete a copy of the "Student Welding Report" listed in Appendix I or provided by your instructor.

VERTICAL WELDS
PRACTICE 4-9

Module 1
Key Indicator 1, 2, 3, 4

Module 2
Key Indicator 1, 2, 3, 4, 7

Module 6
Key Indicator 1, 2
Gas Shielded
Key Indicator 3
Gas Shielded
Key Indicator 4
Self Shielded
Key Indicator 8
Self Shielded
Key Indicator 9

Butt Joint at a 45° Vertical Up Angle

Using a properly set up and adjusted FCA welding machine, proper safety protection, 0.035-in. and/or 0.045-in. (0.9-mm and/or 1.2-mm)-diameter E71T-1 and/or E71T-11 electrodes, and one or more pieces of mild steel plate, 12 in. (305 mm) long and 1/4 in. (6 mm) thick or thinner, you will increase the plate angle gradually as you develop skill until you are making satisfactory welds in the vertical up position, Figure 4.37.

- Start practicing this weld with the plate at a 45° angle.
- Gradually increase the angle of the plate to vertical as skill is gained in welding this joint. A straight stringer bead or slight zigzag will work well on this joint.
- Establish a molten weld pool in the root of the joint.
- Cool, chip, and inspect the weld for uniformity and defects.

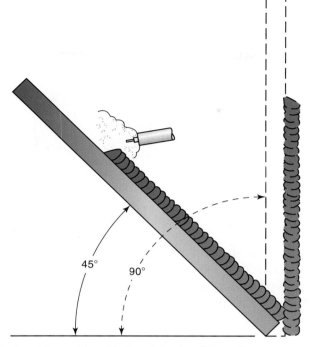

Figure 4.37 Start making welds with the plate at a 45° angle
As your skill develops, increase the angle until the plate is vertical.

It is easier to make a quality weld in the vertical up position if lower settings are used in both the **amperage range** and **voltage range**. This will make the molten weld pool smaller, less fluid, and easier to control. A problem with lower power settings is that the weld bead often can be very convex, Figure 4.38. Faster travel speed and/or slightly wider weave patterns can be used to control the bead shape.

Start at the bottom of the plate and hold the welding gun at a slight upward angle to the plate, Figure 4.39. Brace yourself, lower your hood, and begin to weld. Depending on the machine settings and type of electrode used, you will make a weave pattern.

If the molten weld pool is large and fluid (hot), use a C or J weave pattern to allow a longer time for the molten weld pool to cool, Figure 4.40. Do not make the weave so long or fast that the electrode is allowed to strike the metal ahead of the molten weld pool. If this happens, spatter increases and a spot or zone of incomplete fusion may occur.

A weld that is high and has little or no fusion is too "cold." Changing the welding technique will not correct this problem. The welder must stop welding and make the needed adjustments to the power supply or electrode feeder. Continue to weld along the entire 12-in. (305-mm) length of plate.

Repeat welds with both electrodes until defect-free welds can be consistently made vertically in the 1/4-in. (6-mm)-thick plate. Turn off the

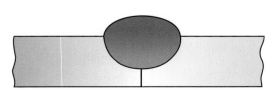
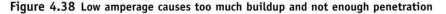

Figure 4.38 Low amperage causes too much buildup and not enough penetration

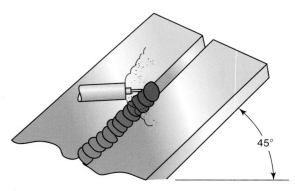

Figure 4.39 45° vertical up

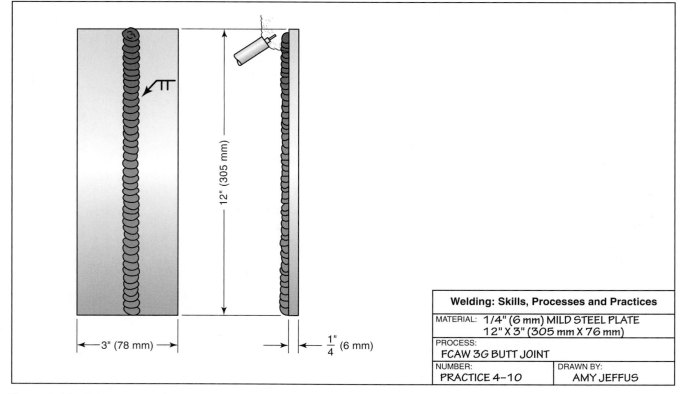

CORRECT
WELD POOL
SIZE AND SHAPE

THIS SHAPE INDICATES
THAT WELD POOL IS
COOLING TOO SLOWLY

Figure 4.40 The shape of the weld pool can indicate the temperature of the surrounding base metal

Module 1
Key Indicator 1, 2, 3, 4

Module 2
Key Indicator 1, 2, 3, 4, 7

Module 6
Gas Shielded
Key Indicator 6
Self Shielded
Key Indicator 11
This practice addresses the
"Vertical" portion of the
all-position requirement.

welding machine and shielding gas and clean up your work area when you are finished welding.

Complete a copy of the "Student Welding Report" listed in Appendix I or provided by your instructor.

PRACTICE 4-10

Butt Joint 3G

Using a properly set up and adjusted FCA welding machine, proper safety protection, E71T-1 and/or E71T-11 electrodes of diameter 0.035 in. and/or 0.045 in. (0.9 mm and/or 1.2 mm), and one or more pieces of mild

← 3" (78 mm) →

12" (305 mm)

$\frac{1}{4}$" (6 mm)

Welding: Skills, Processes and Practices

MATERIAL:	1/4" (6 mm) MILD STEEL PLATE 12" X 3" (305 mm X 76 mm)	
PROCESS: FCAW 3G BUTT JOINT		
NUMBER: PRACTICE 4-10	DRAWN BY: AMY JEFFUS	

Figure 4.41 FCAW 3G butt joint, 1/4 in. mild steel

steel plate, 12 in. (305 mm) long and 1/4 in. (6 mm) thick or thinner, you will make a groove weld in the vertical position, Figure 4.41.

Following the same instructions for the assembly and welding procedure outlined in Practice 4-9, repeat with both classifications of electrodes until defect-free welds can be consistently made in the 1/4-in. (6-mm)-thick plate. Turn off the welding machine and shielding gas and clean up your work area when you are finished welding.

Complete a copy of the "Student Welding Report" listed in Appendix I or provided by your instructor.

PRACTICE 4-11

Butt Joint 3G

Use a properly set up and adjusted FCA welding machine; proper safety protection; E71T-1 and/or E71T-11 electrodes of diameter 0.035 in. and/or 0.045 in. (0.9 mm and/or 1.2 mm); one or more pieces of mild steel plate, beveled, 12 in. (305 mm) long and 3/8 in. (9.5 mm) thick; and a backing strip 14 in. (355 mm) long, 1 in. (25 mm) wide, and 1/4 in. (6 mm) thick. You will make a groove weld in the vertical position.

Following the same instructions for assembly and welding procedure as outlined in Practice 4-9, repeat with both classifications of electrodes until defect-free welds can consistently be made. Turn off the welding machine and shielding gas and clean up your work area when you are finished welding.

Complete a copy of the "Student Welding Report" listed in Appendix I or provided by your instructor.

PRACTICE 4-12

Butt Joint 3G 100% to Be Tested

Use a properly set up and adjusted FCA welding machine; proper safety protection; E71T-1 and/or E71T-11 electrodes of diameter 0.035 in. and/or 0.045 in. (0.9 mm and/or 1.2 mm); one or more pieces of mild steel plate, beveled, 12 in. (305 mm) long and 3/8 in. (9.5 mm) thick; and a backing strip 14 in. (355 mm) long, 1 in. (25 mm) wide, and 1/4 in. (6 mm) thick. You will make a groove weld in the vertical position, Figure 4.42.

Following the same instructions for the assembly and welding procedure outlined in Practice 4-9, repeat the weld until you can use each electrode type to make welds with 100% penetration that will pass a bend test. Turn off the welding machine and shielding gas and clean up your work area when you are finished welding.

Complete a copy of the "Student Welding Report" listed in Appendix I or provided by your instructor.

PRACTICE 4-13

Butt Joint 3G 100% to Be Tested

Use a properly set up and adjusted FCA welding machine; proper safety protection; E71T-1 and/or E71T-11 electrodes of diameter 0.045 in. and/or through 1/16 in. (0.9 mm and/or through 1.6 mm); one or more pieces of mild steel plate, beveled, 7 in. (178 mm) long and

Module 1
Key Indicator 1, 2, 3, 4

Module 2
Key Indicator 1, 2, 3, 4, 7

Module 6
Gas Shielded
Key Indicator 6
Self Shielded
Key Indicator 11

Module 1
Key Indicator 1, 2, 3, 4

Module 2
Key Indicator 1, 2, 3, 4, 7

Module 6
Gas Shielded
Key Indicator 6
Self Shielded
Key Indicator 11

Module 9
Key Indicator 1, 2
This practice addresses the "Vertical" component of the all-position requirement of 6 and 11.

Module 1
Key Indicator 1, 2, 3, 4

Module 2
Key Indicator 1, 2, 3, 4, 7

Module 6
Gas Shielded
Key Indicator 6
Self Shielded
Key Indicator 11

Module 9
Key Indicator 1, 2
This practice addresses the "Vertical" portion of the all-position requirement of 6 and 11.

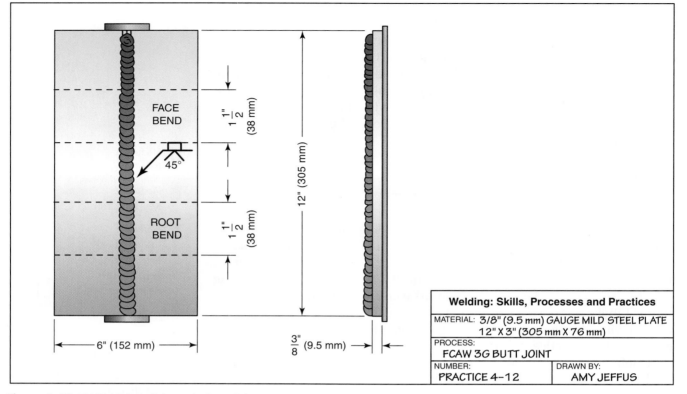

Figure 4.42 FCAW 3G butt joint, 3/8 in. mild steel

3/4 in. (19 mm) thick or thicker; and a backing strip 9 in. (230 mm) long, 1 in. (25 mm) wide, and 1/4 in. (6 mm) thick. You will make a groove weld in the vertical position, Figure 4.43.

Following the same instructions for the assembly and welding procedure outlined in Practice 4-9, repeat the weld until you can use each electrode type to make welds with 100% penetration that will pass a bend test. Turn off the welding machine and shielding gas and clean up your work area when you are finished welding.

Complete a copy of the "Student Welding Report" listed in Appendix I or provided by your instructor.

Module 1
Key Indicator 1, 2, 3, 4

Module 2
Key Indicator 1, 2, 3, 4, 7

Module 6
Key Indicator 1, 2
Gas Shielded
Key Indicator 3
Gas Shielded
Key Indicator 4
Self Shielded
Key Indicator 8
Self Shielded
Key Indicator 9
Self Shielded
Key Indicator 10
Self Shielded
Key Indicator 11
Self Shielded
Key Indicator 12

Module 9
Key Indicator 1, 2

PRACTICE 4-14

Fillet Weld Joint at a 45° Vertical Up Angle

Using a properly set up and adjusted FCA welding machine, proper safety protection, E71T-1 and/or E71T-11 electrodes of diameter 0.035 in. and/or 0.045 in. (0.9 mm and/or 1.2 mm), and one or more pieces of mild steel plate, 12 in. (305 mm) long and 3/8 in. (9.5 mm) thick, you will increase the plate angle gradually as you develop skill until you are making satisfactory welds in the vertical up position, Figure 4.44.

Tack weld the metal pieces together and brace them in position. Check to see that you have free movement for your gun along the entire joint to prevent stopping and restarting during the weld. Avoiding stops and starts both speeds up the welding time and eliminates discontinuities.

It is easier to make a quality weld in the vertical up position if both the amperage and voltage are set at the lower end of their ranges. This

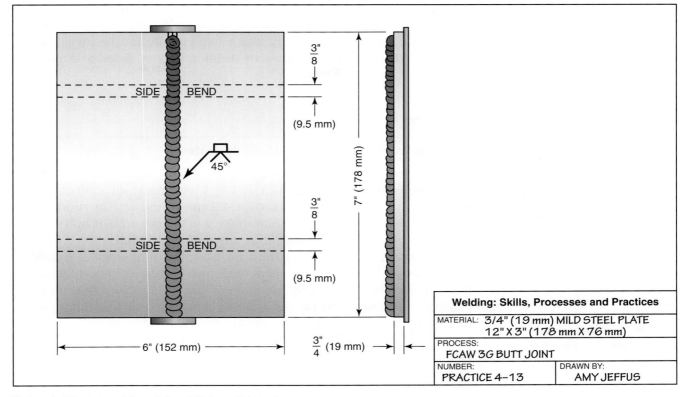

Welding: Skills, Processes and Practices

MATERIAL: 3/4" (19 mm) MILD STEEL PLATE 12" X 3" (178 mm X 76 mm)	
PROCESS: FCAW 3G BUTT JOINT	
NUMBER: PRACTICE 4-13	DRAWN BY: AMY JEFFUS

Figure 4.43 FCAW 3G butt joint, 3/4 in. mild steel

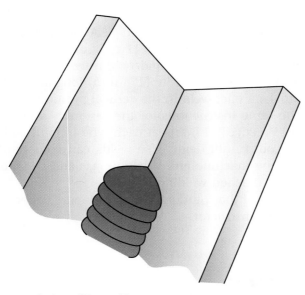

Figure 4.44 45° vertical up fillet weld

will make the molten weld pool smaller, less fluid, and easier to control. A problem with the lower power settings is that the weld bead often is very convex. A convex face on a weld bead often makes it more difficult to remove the slag along the toe of the weld.

The weave pattern should allow for adequate fusion on both edges of the joint. Watch the edges to be sure that they are being melted so that adequate fusion and penetration occur.

Repeat the weld with each electrode type until defect-free welds can consistently be made vertically. Turn off the welding machine and shielding gas and clean up your work area when you are finished welding.

Complete a copy of the "Student Welding Report" listed in Appendix I or provided by your instructor.

Module 1
Key Indicator 1, 2, 3, 4

Module 2
Key Indicator 1, 2, 3, 4, 7

Module 6
Gas Shielded
Key Indicator 5
Self Shielded
Key Indicator 10

Module 9
Key Indicator 1, 2
This practice addresses the
"Vertical" position component
of the all-position requirement.

PRACTICE 4-15

Lap Joint and Tee Joint 3F 100% to Be Tested

Use a properly set up and adjusted FCA welding machine; proper safety protection; E71T-1 and/or E71T-11 electrodes of diameter 0.035 in. and/or 0.045 in. (0.9 mm and/or 1.2 mm); and one or more pieces of mild steel plate, beveled, 12 in. (305 mm) long and 3/8 in. (9.5 mm) thick. You will make a fillet weld in the vertical position.

Following the same instructions for assembly and welding procedure as outlined in Practice 4-14, repeat each type of joint with both classifications of electrodes until you can make welds with 100% penetration that will pass the test. Turn off the welding machine and shielding gas and clean up your work area when you are finished welding.

Complete a copy of the "Student Welding Report" listed in Appendix I or provided by your instructor.

Module 1
Key Indicator 1, 2, 3, 4

Module 2
Key Indicator 1, 2, 3, 4, 7

Module 6
Gas Shielded
Key Indicator 5
Self Shielded
Key Indicator 10
This practice addresses the
"Vertical" component of the
all-position requirement for
5 and 10.

PRACTICE 4-16

Tee Joint 3F

Using a properly set up and adjusted FCA welding machine, proper safety protection, E71T-1 and/or E71T-11 electrodes of diameter 0.045 in. and/or through 1/16 in. (0.9 mm and/or through 1.6 mm), and one or more pieces of mild steel plate, 7 in. (178 mm) long and 3/4 in. (19 mm) thick, you will make a fillet weld in the vertical position.

Following the same instructions for the assembly and welding procedure outlined in Practice 4-14, repeat each type of joint with both classifications of electrodes until defect-free welds can consistently be made. Turn off the welding machine and shielding gas and clean up your work area when you are finished welding.

Complete a copy of the "Student Welding Report" listed in Appendix I or provided by your instructor.

Module 1
Key Indicator 1, 2, 3, 4

Module 2
Key Indicator 1, 2, 3, 4, 7

Module 6
Gas Shielded
Key Indicator 5
Self Shielded
Key Indicator 10
This practice addresses the
"Horizontal" component of the
all-position requirement for
5 and 10.

HORIZONTAL WELDS

PRACTICE 4-17

Lap Joint and Tee Joint 2F

Use a properly set up and adjusted FCA welding machine; proper safety protection; E70T-1 and/or E71T-11 (or E71T-1 and/or E70T-5) electrodes of diameter 0.035 in. and/or 0.045 in. (0.9 mm and/or 1.2 mm); and one or more pieces of mild steel plate, beveled, 12 in. (305 mm) long and 3/8 in. (9.5 mm) thick. You will make a fillet weld in the horizontal position, Figure 4.45A and Figure 4.45B.

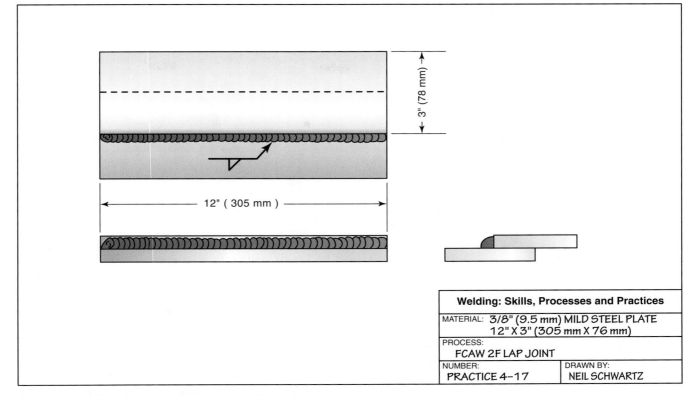

Figure 4.45 (A) FCAW 2F lap joint, 3/8 in. mild steel

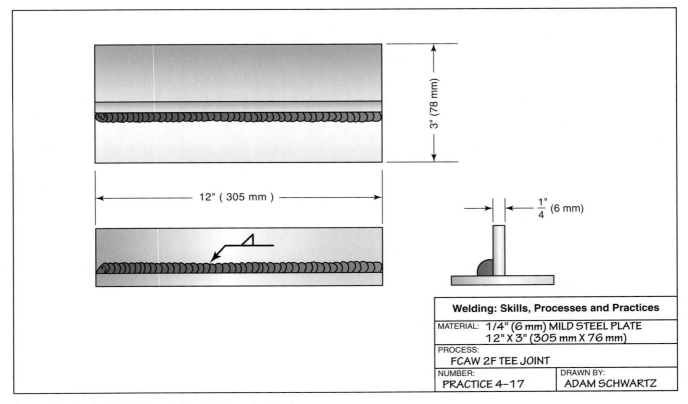

Figure 4.45 (B) FCAW 2F tee joint, 1/4 in. mild steel

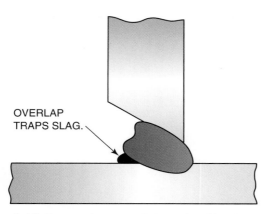

Figure 4.46 Slag can be trapped along the side of the root pass

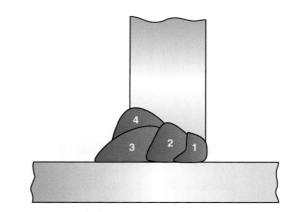

Figure 4.47 FCAW weld bead positions for a 100% penetration grooved tee joint

The root weld must be kept small so that its contour can be controlled. Too large a root pass can trap slag under overlap along the lower edge of the weld, Figure 4.46. Clean each pass thoroughly before the weld bead is started. Follow the weld bead sequence shown in Figure 4.47. Use stringer beads rather than weave beads in order to eliminate overlap. Keeping all of the weld beads small will help control their contour.

Repeat each type of joint with both classifications of electrodes until defect-free welds can consistently be made. Turn off the welding machine and shielding gas and clean up your work area when you are finished welding.

Complete a copy of the "Student Welding Report" listed in Appendix I or provided by your instructor.

Module 1
Key Indicator 1, 2, 3, 4

Module 2
Key Indicator 1, 2, 3, 4, 7

Module 6
Gas Shielded
Key Indicator 5
Self Shielded
Key Indicator 10
This practice addresses the
"Horizontal" component of the
all-position requirement for
5 and 10.

PRACTICE 4-18

Tee Joint 2F

Use a properly set up and adjusted FCA welding machine, proper safety protection, E70T-1 and/or E71T-11 (or E71T-1 and/or E70T-5) electrodes of diameter 0.045 in. and/or through 1/16 in. (0.9 mm and/or through 1.6 mm), and one or more pieces of mild steel, beveled, 7 in. (178 mm) long and 3/4 in. (19 mm) thick or thicker. You will make a fillet weld in the horizontal position.

Following the same instructions for the assembly and welding procedure outlined in Practice 4-17, repeat each type of joint with both classifications of electrodes until defect-free welds can consistently be made. Turn off the welding machine and shielding gas and clean up your work area when you are finished welding.

Complete a copy of the "Student Welding Report" listed in Appendix I or provided by your instructor.

Module 1
Key Indicator 1, 2, 3, 4

Module 2
Key Indicator 1, 2, 3, 4, 7

Module 6
Gas Shielded
Key Indicator 6
Self Shielded
Key Indicator 11
This practice addresses the
"Horizontal" position portion of
the all-position requirement
of 6 and 11.

PRACTICE 4-19

Butt Joint 2G

Use a properly set up and adjusted FCA welding machine, proper safety protection, E70T-1 and/or E71T-11 (or E71T-1 and/or E70T-5 electrodes) of diameter 0.035 in. and/or 0.045 in. (0.9 mm and/or 1.2 mm),

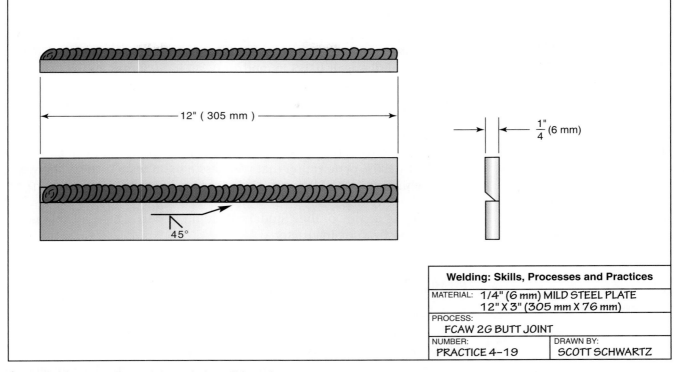

Welding: Skills, Processes and Practices

MATERIAL: 1/4" (6 mm) MILD STEEL PLATE 12" X 3" (305 mm X 76 mm)	
PROCESS: FCAW 2G BUTT JOINT	
NUMBER: PRACTICE 4-19	DRAWN BY: SCOTT SCHWARTZ

Figure 4.48 FCAW 2G butt joint, 1/4 in. mild steel

and one or more pieces of mild steel plate, 12 in. (305 mm) long and 1/4 in. (6 mm) thick or thinner. You will make a groove weld in the horizontal position, Figure 4.48.

Repeat with both classifications of electrodes until defect-free welds can consistently be made in the 1/4-in. (6-mm)-thick plate. Turn off the welding machine and shielding gas and clean up your work area when you are finished welding.

Complete a copy of the "Student Welding Report" listed in Appendix I or provided by your instructor.

PRACTICE 4-20

Butt Joint 2G

Use a properly set up and adjusted FCA welding machine; proper safety protection; E70T-1 and/or E71T-11 or 5 (or E71T-1 and/or E70T-5) electrodes of diameter 0.035 in. and/or 0.045 in. (0.9 mm and/or 1.2 mm); one or more pieces of mild steel plate, beveled, 12 in. (305 mm) long and 3/8 in. (9.5 mm) thick; and a backing strip 14 in. (355 mm) long, 1 in. (25 mm) wide, and 1/4 in. (6 mm) thick. You will make a groove weld in the horizontal position.

Repeat with both classifications of electrodes until defect-free welds can consistently be made. Turn off the welding machine and shielding gas and clean up your work area when you are finished welding.

Complete a copy of the "Student Welding Report" listed in Appendix I or provided by your instructor.

Module 1
Key Indicator 1, 2, 3, 4

Module 2
Key Indicator 1, 2, 3, 4, 7

Module 6
Gas Shielded
Key Indicator 6
Self Shielded
Key Indicator 11
This practice addresses the "Horizontal" position portion of the all-position requirement of 6 and 11.

PRACTICE 4-21

Butt Joint 2G 100% to Be Tested

Use a properly set up and adjusted FCA welding machine; proper safety protection; E70T-1 and/or E71T-11 or 5 (or E71T-1 and/or E70T-5) electrodes of diameter 0.035 in. and/or 0.045 in. (0.9 mm and/or 1.2 mm); one or more pieces of mild steel plate, beveled, 12 in. (305 mm) long and 3/8 in. (9.5 mm) thick; and a backing strip 14 in. (355 mm) long, 1 in. (25 mm) wide and 1/4 in. (6 mm) thick. You will make a groove weld in the horizontal position.

Repeat the weld using each electrode classification until you can make welds with 100% penetration that will pass a bend test. Turn off the welding machine and shielding gas and clean up your work area when you are finished welding.

Complete a copy of the "Student Welding Report" listed in Appendix I or provided by your instructor.

PRACTICE 4-22

Butt Joint 2G

Use a properly set up and adjusted FCA welding machine; proper safety protection; E70T-1 and/or E71T-11 electrodes (or E71T-1 and/or E70T-5) of diameter 0.035 in. and/or through 1/16 in. (0.9 mm and/or through 1.6 mm); one or more pieces of mild steel plate, beveled, 7 in. (178 mm) long and 3/4 in. (19 mm) thick or thicker; and a backing strip 9 in. (230 mm) long, 1 in. (25 mm) wide, and 1/4 in. (6 mm) thick. You will make a groove weld in the horizontal position, Figure 4.49.

Repeat with both classifications of electrodes until defect-free welds can consistently be made. Turn off the welding machine and shielding gas and clean up your work area when you are finished welding.

Complete a copy of the "Student Welding Report" listed in Appendix I or provided by your instructor.

PRACTICE 4-23

Butt Joint 2G 100% to Be Tested

Use a properly set up and adjusted FCA welding machine; proper safety protection; E70T-1 and/or E71T-11 electrodes (or E71T-1 and/or E70T-5) of diameter 0.045 in. and/or through 1/16 in. (0.9 mm and/or through 1.6 mm); one or more pieces of mild steel plate, beveled, 7 in. (178 mm) long and 3/4 in. (19 mm) thick or thicker; and a backing strip 9 in. (230 mm) long, 1 in. (25 mm) wide, and 1/4 in. (6 mm) thick. You will make a groove weld in the horizontal position.

Repeat the weld until you can use each electrode type to make welds with 100% penetration that will pass a bend test. Turn off the welding machine and shielding gas and clean up your work area when you are finished welding.

Complete a copy of the "Student Welding Report" listed in Appendix I or provided by your instructor.

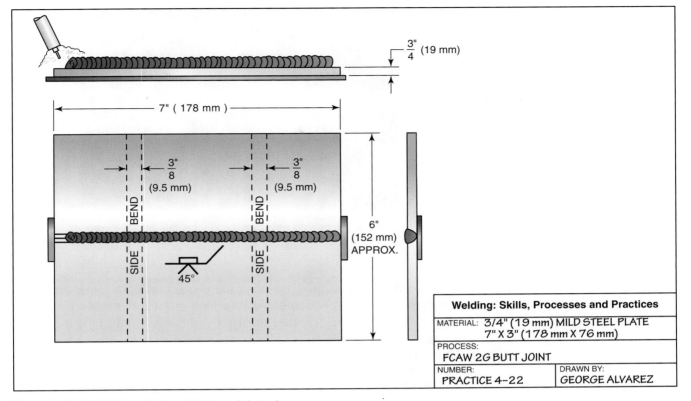

Welding: Skills, Processes and Practices

MATERIAL:	3/4" (19 mm) MILD STEEL PLATE 7" X 3" (178 mm X 76 mm)	
PROCESS: FCAW 2G BUTT JOINT		
NUMBER: PRACTICE 4-22	DRAWN BY: GEORGE ALVAREZ	

Figure 4.49 FCAW 2G butt joint, 3/4 in. mild steel

OVERHEAD-POSITION WELDS
PRACTICE 4-24

Butt Joint 4G

Using a properly set up and adjusted FCA welding machine, proper safety protection, E71T-1 and/or E71T-11 electrodes of diameter 0.035 in. and/or 0.045 in. (0.9 mm and/or 1.2 mm), and one or more pieces of mild steel plate, 12 in. (305 mm) long and 1/4 in. (6 mm) thick or thinner, you will make a groove weld in the overhead position.

The molten weld pool should be kept as small as possible for easier control. A small molten weld pool can be achieved by using lower current, faster traveling speeds, and settings.

Lower current settings require closer control of gun manipulation to ensure that the electrode is fed into the molten weld pool just behind the leading edge. The low power will cause overlap and more spatter if this electrode-to-molten weld pool contact position is not closely maintained.

Faster travel speeds allow the welder to maintain a high production rate even if multiple passes are required to complete the weld. Weld penetration into the base metal at the start of the bead can be obtained by using a slow start or quickly reversing the weld direction. Both the slow start and reversal of weld direction put more heat into the weld start to increase penetration. The higher speed also reduces the amount of weld distortion by reducing the amount of time that heat is applied to a joint.

Module 1
Key Indicator 1, 2, 3, 4

Module 2
Key Indicator 1, 2, 3, 4, 7

Module 6
Gas Shielded
Key Indicator 6
Self Shielded
Key Indicator 11
This practice addresses the "Overhead" position portion of the all-position requirement of 6 and 11.

Figure 4.50 Hold the gun so that weld spatter will not fall onto the gun
Source: Courtesy of Larry Jeffus

For overhead welding, extra personal protection is required to reduce the danger of burns. Leather sleeves or leather jackets should be worn.

Much of the spatter created during overhead welding falls into or on the nozzle and contact tube. The contact tube may short out to the gas nozzle. The shorted gas nozzle may arc to the work, causing damage both to the nozzle and to the plate. To control the amount of spatter, a longer stickout and/or a sharper gun-to-plate angle is required to allow most of the spatter to fall clear of the gun or nozzle, Figure 4.50.

Make several short weld beads using various techniques to establish the method that is most successful and most comfortable for you. After each weld, stop and evaluate it before making a change. When you have decided on the technique to be used, make a welded stringer bead that is 12 in. (305 mm) long.

Repeat with both classifications of electrodes until defect-free welds can consistently be made in the 1/4-in. (6-mm)-thick plate. Turn off the welding machine and shielding gas and clean up your work area when you are finished welding.

Complete a copy of the "Student Welding Report" listed in Appendix I or provided by your instructor.

Module 1
Key Indicator 1, 2, 3, 4

Module 2
Key Indicator 1, 2, 3, 4, 7

Module 6
Gas Shielded
Key Indicator 6
Self Shielded
Key Indicator 11

Module 9
Key Indicator 1, 2
This practice addresses the "Overhead" position portion of the all-position requirement of 6 and 11.

PRACTICE 4-25

Butt Joint 4G 100% to Be Tested

Using a properly set up and adjusted FCA welding machine, proper safety protection, E71T-1 and/or E71T-11 electrodes of diameter 0.035 in. and/or 0.045 in. (0.9 mm and/or 1.2 mm), and one or more pieces of mild steel plate, 12 in. (305 mm) long and 1/4 in. (6 mm) thick, you will make a groove weld in the overhead position.

Following the same instructions for the assembly and welding procedure outlined in Practice 4-24, repeat the weld until you can use each electrode type to make welds with 100% penetration that will pass a bend test. Turn off the welding machine and shielding gas and clean up your work area when you are finished welding.

Complete a copy of the "Student Welding Report" listed in Appendix I or provided by your instructor.

PRACTICE 4-26

Butt Joint 4G

Use a properly set up and adjusted FCA welding machine; proper safety protection; E71T-1 and/or E71T-11 electrodes of diameter 0.035 in. and/or 0.045 in. (0.9 mm and/or 1.2 mm); one or more pieces of mild steel plate, beveled, 12 in. (305 mm) long and 3/8 in. (9.5 mm) thick; and a backing strip 14 in. (355 mm) long, 1 in. (25 mm) wide, and 1/4 in. (6 mm) thick. You will make a groove weld in the overhead position, Figure 4.51.

Following the same instructions for the assembly and welding procedure outlined in Practice 4-24, repeat with both classifications of electrodes until defect-free welds can consistently be made. Turn off the welding machine and shielding gas and clean up your work area when you are finished welding.

Complete a copy of the "Student Welding Report" listed in Appendix I or provided by your instructor.

PRACTICE 4-27

Butt Joint 4G 100% to Be Tested

Use a properly set up and adjusted FCA welding machine; proper safety protection; E71T-1 and/or E71T-11 electrodes of diameter 0.035 in. and/or 0.045 in. (0.9 mm and/or 1.2 mm); one or more pieces of mild steel plate, beveled, 12 in. (305 mm) long and 3/8 in. (9.5 mm) thick; and a backing strip 14 in. (355 mm) long, 1 in. (25 mm) wide, and 1/4 in. (6 mm) thick. You will make a groove weld in the overhead position.

Module 1
Key Indicator 1, 2, 3, 4

Module 2
Key Indicator 1, 2, 3, 4, 7

Module 6
Gas Shielded
Key Indicator 6
Self Shielded
Key Indicator 11
This practice addresses the "Overhead" position portion of the all-position requirement of 6 and 11.

Module 1
Key Indicator 1, 2, 3, 4

Module 2
Key Indicator 1, 2, 3, 4, 7

Module 6
Gas Shielded
Key Indicator 6
Self Shielded
Key Indicator 11

Module 9
Key Indicator 1, 2
This practice addresses the "Overhead" position portion of the all-position requirement of 6 and 11.

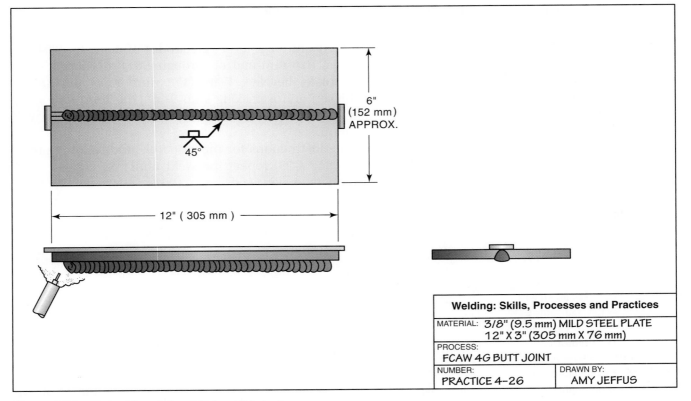

6"
(152 mm)
APPROX.

45°

12" (305 mm)

Welding: Skills, Processes and Practices

MATERIAL:	3/8" (9.5 mm) MILD STEEL PLATE 12" X 3" (305 mm X 76 mm)
PROCESS:	FCAW 4G BUTT JOINT
NUMBER: PRACTICE 4-26	DRAWN BY: AMY JEFFUS

Figure 4.51 FCAW 4G butt joint, 3/8 in. mild steel

Following the same instructions for the assembly and welding procedure outlined in Practice 4-24, repeat the weld until you can use each electrode type to make welds with 100% penetration that will pass a bend test. Turn off the welding machine and shielding gas and clean up your work area when you are finished welding.

Complete a copy of the "Student Welding Report" listed in Appendix I or provided by your instructor.

PRACTICE 4-28

Butt Joint 4G

Use a properly set up and adjusted FCA welding machine; proper safety protection; E71T-1 and/or E71T-11 electrodes of diameter 0.045 in. and/or through 1/16 in. (0.9 mm and/or through 1.6 mm); one or more pieces of mild steel plate, beveled, 7 in. (178 mm) long and 3/4 in. (19 mm) thick or thicker; and a backing strip 9 in. (230 mm) long, 1 in. (25 mm) wide, and 1/4 in. (6 mm) thick. You will make a groove weld in the overhead position.

Following the same instructions for the assembly and welding procedure outlined in Practice 4-24, repeat with both classifications of electrodes until defect-free welds can consistently be made. Turn off the welding machine and shielding gas and clean up your work area when you are finished welding.

Complete a copy of the "Student Welding Report" listed in Appendix I or provided by your instructor.

PRACTICE 4-29

Butt Joint 4G 100% to Be Tested

Use a properly set up and adjusted FCA welding machine; proper safety protection; E71T-1 and/or E71T-11 electrodes of diameter 0.045 in. and/or through 1/16 in. (0.9 mm and/or through 1.6 mm); one or more pieces of mild steel plate, beveled, 7 in. (178 mm) long and 3/4 in. (19 mm) thick or thicker; and a backing strip 9 in. (230 mm) long, 1 in. (25 mm) wide, and 1/4 in. (6 mm) thick. You will make a groove weld in the overhead position.

Following the same instructions for the assembly and welding procedure outlined in Practice 4-24, repeat the weld until you can use each electrode classification to make welds with 100% penetration that will pass a bend test. Turn off the welding machine and shielding gas and clean up your work area when you are finished welding.

Complete a copy of the "Student Welding Report" listed in Appendix I or provided by your instructor.

PRACTICE 4-30

Lap Joint and Tee Joint 4F

Use a properly set up and adjusted FCA welding machine; proper safety protection; E71T-1 and/or E71T-11 electrodes of diameter 0.035 in. and/or 0.045 in. (0.9 mm and/or 1.2 mm); and one or more pieces of mild steel plate, beveled, 12 in. (305 mm) long and 3/4 in. (19 mm) thick. You will make a fillet weld in the overhead position, Figure 4.52A and Figure 4.52B.

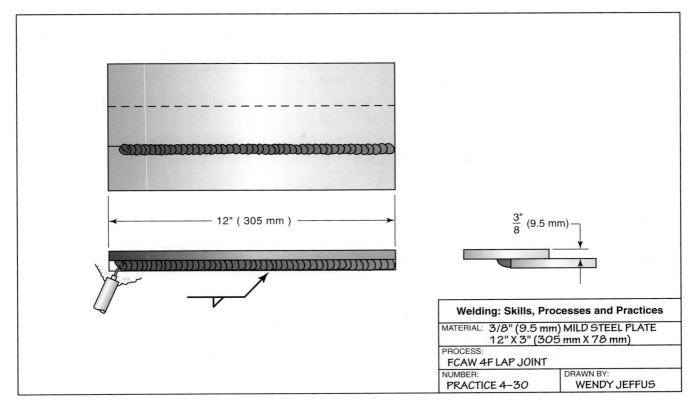

Welding: Skills, Processes and Practices

MATERIAL: 3/8" (9.5 mm) MILD STEEL PLATE 12" X 3" (305 mm X 78 mm)	
PROCESS: FCAW 4F LAP JOINT	
NUMBER: PRACTICE 4-30	DRAWN BY: WENDY JEFFUS

Figure 4.52 (A) FCAW 4F lap joint, 3/8 in. mild steel

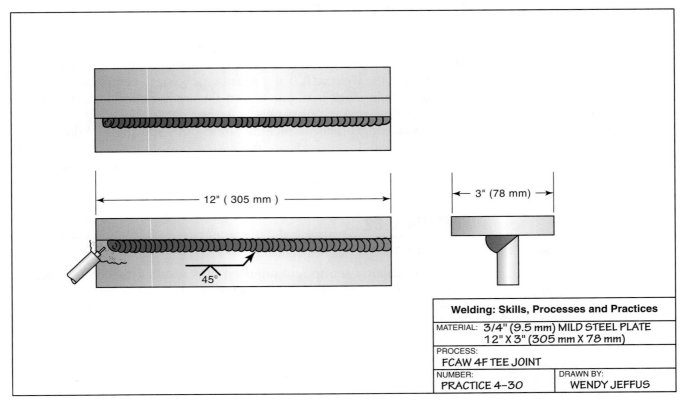

Welding: Skills, Processes and Practices

MATERIAL: 3/4" (9.5 mm) MILD STEEL PLATE 12" X 3" (305 mm X 78 mm)	
PROCESS: FCAW 4F TEE JOINT	
NUMBER: PRACTICE 4-30	DRAWN BY: WENDY JEFFUS

Figure 4.52 (B) FCAW 4F tee joint, 3/4 in. mild steel

Following the same instructions for the assembly and welding procedure outlined in Practice 4-24, repeat each type of joint with both classifications of electrodes until defect-free welds can consistently be made. Turn off the welding machine and shielding gas and clean up your work area when you are finished welding.

Complete a copy of the "Student Welding Report" listed in Appendix I or provided by your instructor.

PRACTICE 4-31

Tee Joint 4F 100% to Be Tested

Module 1
Key Indicator 1, 2, 3, 4

Module 2
Key Indicator 1, 2, 3, 4, 7

Module 6
Gas Shielded
Key Indicator 5
Self Shielded
Key Indicator 10

Module 9
Key Indicator 1, 2
This practice addresses the "Overhead" component of the all-position requirement for 5 and 10.

Use a properly set up and adjusted FCA welding machine; proper safety protection; E71T-1 and/or E71T-11 electrodes of diameter 0.045 in. and/or through 1/16 in. (0.9-mm and/or through 1.6-mm); and one or more pieces of mild steel plate, beveled, 7 in. (178 mm) long and 3/4 in. (19 mm) thick or thicker. You will make a fillet weld in the overhead position.

Following the same instructions for the assembly and welding procedure as outlined in Practice 4-24, repeat the weld until you can use each electrode classification to make welds with 100% penetration that will pass the bend test. Turn off the welding machine and shielding gas and clean up your work area when you are finished welding.

Complete a copy of the "Student Welding Report" listed in Appendix I or provided by your instructor.

THIN-GAUGE WELDING

The introduction of small electrode diameters has allowed FCA welding to be used on thin sheet metal. Usually these welds will be a fillet type, the easiest weld to make on thin stock. An effort should be made when possible to design the weld so it is not a butt-type joint. A common use for FCA welding on thin stock is to join it to a thicker member, Figure 4.53. This type of weld is used to put panels in frames.

The following practices include some butt-type joints. You will find that the vertical down welds are the easiest ones to make. If it is possible to position the weldment for a vertical down weld, production speeds for butt joints can be increased.

PRACTICE 4-32

Butt Joint 1G

Module 1
Key Indicator 1, 2, 3, 4

Module 2
Key Indicator 1, 2, 3, 4, 7

Module 6
Gas Shielded
Key Indicator 6
Self Shielded
Key Indicator 11

Use a properly set up and adjusted FCA welding machine, Table 4.3; proper safety protection; E71T-1 and/or E70T-5 or E70T-1 and/or E71T-11 electrodes of diameter 0.030 in. and/or 0.035 in. (0.8 mm and/or 0.9 mm); and one or more pieces of mild steel sheet, 12 in. (305 mm) long

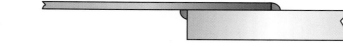

Figure 4.53 FCA welding of thin to thick metal

Table 4.3 FCA Welding Parameters for Use if Specific Settings Are Unavailable from Electrode Manufacturer

Electrode		Welding Power			Shielding Gas		Base Metal	
Type	Size	Amps	Wire-feed Speed, ipm (cm/min)	Volts	Type	Flow	Type	Thick
E70T-1 E71T-1	0.030 in. (0.8 mm)	40 to 145	90 to 340 (228 to 864)	20 to 27	None	n/a	Low-carbon steel	16 gauge to 18 gauge
E70T-1 E71T-1	0.035 in. (0.9 mm)	130 to 200	288 to 576 (732 to 1463)	20 to 28	None	n/a	Low-carbon steel	16 gauge to 18 gauge
E70T-5 E71T-11	0.035 in. (0.9 mm)	90 to 200	190 to 576 (483 to 1463)	16 to 29	57% argon 25% CO_2	35 cfh	Low-carbon steel	16 gauge 18 gauge

and 16-gauge to 18-gauge thick. You will make a butt weld in the flat position, Figure 4.54.

Do not leave a root opening for these welds. Even the slightest opening will result in a burn-through. If a burn-through occurs, the welder can be pulsed off and on so that the hole can be filled. This process will leave a larger than usual buildup. Excessive buildup could be ground off if necessary as part of the postweld cleanup.

Repeat with both classifications of electrodes until defect-free welds can consistently be made. Turn off the welding machine and shielding gas and clean up your work area when you are finished welding.

Complete a copy of the "Student Welding Report" listed in Appendix I provided by your instructor.

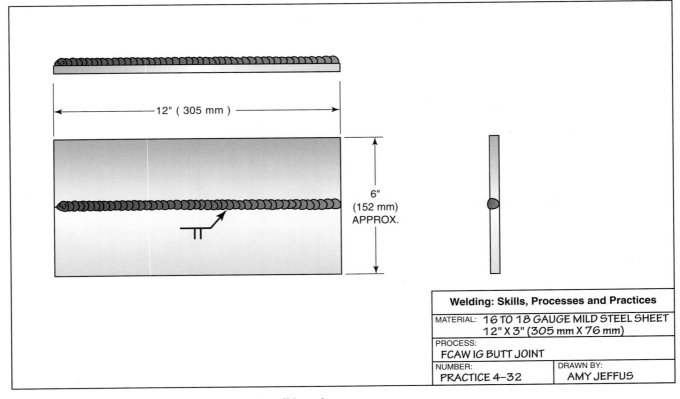

Welding: Skills, Processes and Practices

MATERIAL: *16 TO 18 GAUGE MILD STEEL SHEET 12" X 3" (305 mm X 76 mm)*

PROCESS: *FCAW IG BUTT JOINT*

NUMBER: *PRACTICE 4–32*

DRAWN BY: *AMY JEFFUS*

Figure 4.54 FCAW 1G butt joint, 16- to 18-gauge mild steel

PRACTICE 4-33

Butt Joint 1G 100% to Be Tested

Using a properly set up and adjusted FCA welding machine, proper safety protection, E71T-1 and/or E70T-5 or E70T-1 and/or E70T-11 electrodes of diameter 0.030 in. and/or 0.035 in. (0.8 mm and/or 0.9 mm), and one or more pieces of mild steel sheet, 12 in. (305 mm) long and 16-gauge to 18-gauge thick, you will make a butt weld in the flat position, Figure 4.55.

Following the same instructions for the assembly and welding procedure outlined in Practice 4-32, repeat the weld until you can use each electrode classification to make welds with 100% penetration that will pass a bend test. Turn off the welding machine and shielding gas and clean up your work area when you are finished welding.

Complete a copy of the "Student Welding Report" listed in Appendix I or provided by your instructor.

PRACTICE 4-34

Lap Joint and Tee Joint 1F

Using a properly set up and adjusted FCA welding machine, proper safety protection, E70T-1 and/or E71T-11 or E71T-1 and/or E70T-5 electrodes of diameter 0.030-in. and/or 0.035-in. (0.8-mm and/or 0.9-mm), and one or more pieces of mild steel sheet, 12 in. (305 mm) long and 16-gauge to 18-gauge thick, you will make a fillet weld in the flat position.

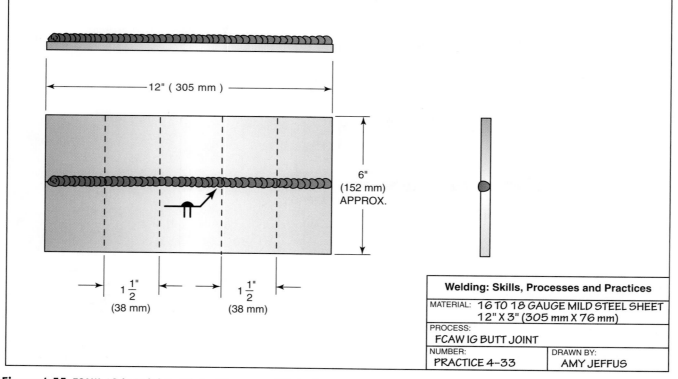

Welding: Skills, Processes and Practices	
MATERIAL: 16 TO 18 GAUGE MILD STEEL SHEET 12" X 3" (305 mm X 76 mm)	
PROCESS: FCAW 1G BUTT JOINT	
NUMBER: PRACTICE 4-33	DRAWN BY: AMY JEFFUS

Figure 4.55 FCAW 1G butt joint, 16- to 18-gauge mild steel

Following the same instructions for the assembly and welding procedure outlined in Practice 4-32, repeat each type of joint with both classifications of electrodes until defect-free welds can consistently be made. Turn off the welding machine and shielding gas and clean up your work area when you are finished welding.

Complete a copy of the "Student Welding Report" listed in Appendix I or provided by your instructor.

PRACTICE 4-35

Lap Joint and Tee Joint 1F 100% to Be Tested

Using a properly set up and adjusted FCA welding machine, proper safety protection, E70T-1 and/or E71T-11 or E71T-1 and/or E70T-5 electrodes of diameter 0.030 in. and/or 0.035 in. (0.8 mm and/or 0.9 mm), and one or more pieces of mild steel sheet, 12 in. (305 mm) long and 16-gauge to 18-gauge thick, you will make a fillet weld in the flat position, Figure 4.56A and Figure 4.56B.

Following the same instructions for the assembly and welding procedure outlined in Practice 4-32, repeat each type of joint with both classifications of electrodes until you can make welds with 100% penetration that will pass the bend test, Figure 4.57A and Figure 4.57B. Turn off the welding machine and shielding gas and clean up your work area when you are finished welding.

Complete a copy of the "Student Welding Report" listed in Appendix I or provided by your instructor.

PRACTICE 4-36

Butt Joint 3G

Using a properly set up and adjusted FCA welding machine, proper safety protection, E71T-1 and/or E71T-11 electrodes of diameter 0.030 in. and/or 0.035 in. (0.8 mm and/or 0.9 mm), and one or more pieces of mild

Module 1
Key Indicator 1, 2, 3, 4

Module 2
Key Indicator 1, 2, 3, 4, 7

Module 6
Gas Shielded
Key Indicator 5
Self Shielded
Key Indicator 10

Module 9
Key Indicator 1, 2

Module 1
Key Indicator 1, 2, 3, 4

Module 2
Key Indicator 1, 2, 3, 4, 7

Module 6
Gas Shielded
Key Indicator 6
Self Shielded
Key Indicator 11
This practice addresses the "Vertical" position portion of the all-position requirement of 6 and 11.

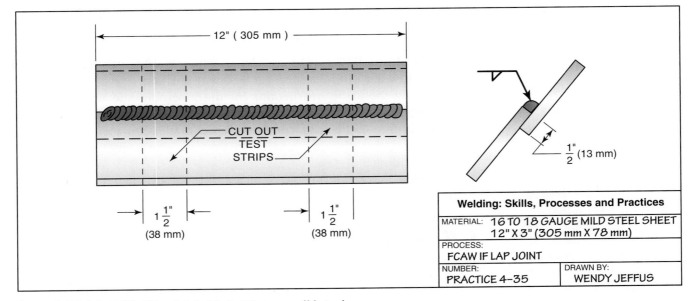

Welding: Skills, Processes and Practices

MATERIAL: *16 TO 18 GAUGE MILD STEEL SHEET 12" X 3" (305 mm X 78 mm)*

PROCESS: *FCAW 1F LAP JOINT*

NUMBER: *PRACTICE 4-35*

DRAWN BY: *WENDY JEFFUS*

Figure 4.56 (A) FCAW 1F lap joint, 16- to 18-gauge mild steel

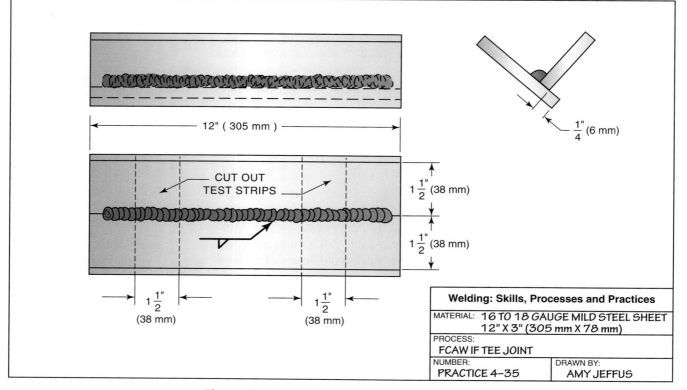

Figure 4.56 (B) FCAW 1F tee joint, 16- to 18-gauge mild steel

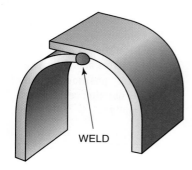

Figure 4.57 (A) 180° bend to test lap weld quality

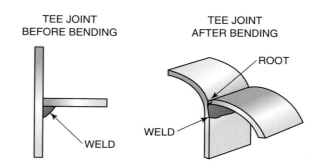

Figure 4.57 (B) Bend the test strip to be sure the weld had good root fusion

Module 1
Key Indicator 1, 2, 3, 4

Module 2
Key Indicator 1, 2, 3, 4, 7

Module 6
Gas Shielded
Key Indicator 5
Self Shielded
Key Indicator 10
This practice addresses the
"Vertical" component of the
all-position requirement
for 5 and 10.

steel sheet, 12 in. (305 mm) long and 16-gauge to 18-gauge thick, you will make a butt weld in the vertical up or down position.

Following the same instructions for the assembly and welding procedure outlined in Practice 4-32, repeat with both classifications of electrodes until defect-free welds can consistently be made. Turn off the welding machine and shielding gas and clean up your work area when you are finished welding.

Complete a copy of the "Student Welding Report" listed in Appendix I or provided by your instructor.

PRACTICE 4-37

Lap Joint and Tee Joint 3F

Using a properly set up and adjusted FCA welding machine, proper safety protection, E71T-1 and/or E71T-11 electrodes of diameter 0.030 in.

and/or 0.035 in. (0.8 mm and/or 0.9 mm), and one or more pieces of mild steel sheet, 12 in. (305 mm) long and 16-gauge to 18- gauge thick, you will make a fillet weld in the vertical up or down position.

Following the same instructions for the assembly and welding procedure outlined in Practice 4-32, repeat each type of joint with both classifications of electrodes until defect-free welds can consistently be made. Turn off the welding machine and shielding gas and clean up your work area when you are finished welding.

Complete a copy of the "Student Welding Report" listed in Appendix I or provided by your instructor.

PRACTICE 4-38

Lap Joint and Tee Joint 3F 100% to Be Tested

Using a properly set up and adjusted FCA welding machine, proper safety protection, E71T-1 and/or E71T-11 electrodes of diameter 0.030 in. and/or 0.035 in. (0.8 mm and/or 0.9 mm) and one or more pieces of mild steel sheet, 12 in. (305 mm) long and 16-gauge to 18-gauge thick, you will make a fillet weld in the vertical up or down position.

Following the same instructions for the assembly and welding procedure outlined in Practice 4-32, repeat each type of joint with both classifications of electrodes until you can make welds with 100% penetration that will pass the test. Turn off the welding machine and shielding gas and clean up your work area when you are finished welding.

Complete a copy of the "Student Welding Report" listed in Appendix I or provided by your instructor.

PRACTICE 4-39

Lap Joint and Tee Joint 2F

Using a properly set up and adjusted FCA welding machine, proper safety protection, E70T-1 and/or E71T-11 electrodes of diameter 0.030 in. and/or 0.035 in. (0.8 mm and/or 0.9 mm) and one or more pieces of mild steel sheet, 12 in. (305 mm) long and 16-gauge to 18-gauge thick, you will make a fillet weld in the horizontal position.

Following the same instructions for the assembly and welding procedure outlined in Practice 4-32, repeat each type of joint with both classifications of electrodes until defect-free welds can consistently be made. Turn off the welding machine and shielding gas and clean up your work area when you are finished welding.

Complete a copy of the "Student Welding Report" listed in Appendix I or provided by your instructor.

PRACTICE 4-40

Lap Joint and Tee Joint 2F 100% to Be Tested

Using a properly set up and adjusted FCA welding machine, proper safety protection, E70T-1 and/or E71T-11 or E71T-1 and/or E70T-5 electrodes of diameter 0.030 in. and/or 0.035 in. (0.8 mm and/or 0.9 mm), and one or more pieces of mild steel sheet, 12 in. (305 mm) long and

Module 1
Key Indicator 1, 2, 3, 4

Module 2
Key Indicator 1, 2, 3, 4, 7

Module 6
Gas Shielded
Key Indicator 5
Self Shielded
Key Indicator 10

Module 9
Key Indicator 1, 2
This practice addresses the "Vertical" component of the all-position requirement for 5 and 10.

Module 1
Key Indicator 1, 2, 3, 4

Module 2
Key Indicator 1, 2, 3, 4, 7

Module 6
Gas Shielded
Key Indicator 5
Self Shielded
Key Indicator 10
This practice addresses the "Horizontal" component of the all-position requirement for 5 and 10.

Module 1
Key Indicator 1, 2, 3, 4

Module 2
Key Indicator 1, 2, 3, 4, 7

Module 6
Gas Shielded
Key Indicator 5
Self Shielded
Key Indicator 10

Module 9
Key Indicator 1, 2
This practice addresses the "Horizontal" component of the all-position requirement for 5 and 10.

16-gauge to 18-gauge thick, you will make a fillet weld in the horizontal position.

Following the same instructions for the assembly and welding procedure outlined in Practice 4-32, repeat each type of joint with both classifications of electrodes until you can make welds with 100% penetration that will pass the bend test. Turn off the welding machine and shielding gas and clean up your work area when you are finished welding.

Complete a copy of the "Student Welding Report" listed in Appendix I or provided by your instructor.

PRACTICE 4-41

Butt Joint 2G

Using a properly set up and adjusted FCA welding machine, proper safety protection, E70T-1 and/or E71T-11 or E71T-1 and/or E70T-5 electrodes of diameter 0.030 in. and/or 0.035 in. (0.8 mm and/or 0.9 mm), and one or more pieces of mild steel sheet, 12 in. (305 mm) long and 16-gauge to 18-gauge thick, you will make a butt weld in the horizontal position.

Following the same instructions for the assembly and welding procedure outlined in Practice 4-32, repeat with both classifications of electrodes until defect-free welds can consistently be made. Turn off the welding machine and shielding gas and clean up your work area when you are finished welding.

Complete a copy of the "Student Welding Report" listed in Appendix I or provided by your instructor.

PRACTICE 4-42

Butt Joint 2G 100% to Be Tested

Using a properly set up and adjusted FCA welding machine, proper safety protection, E70T-1 and/or E71T-11 or E71T-1 and/or E70T-5 electrodes of diameter 0.030 in. and/or 0.035 in. (0.8 mm and/or 0.9 mm), and one or more pieces of mild steel sheet, 12 in. (305 mm) long and 16-gauge to 18-gauge thick, you will make a butt weld in the horizontal position.

Following the same instructions for the assembly and welding procedure outlined in Practice 4-32, repeat the weld until you can use each electrode classification to make welds with 100% penetration that will pass a bend test. Turn off the welding machine and shielding gas and clean up your work area when you are finished welding.

Complete a copy of the "Student Welding Report" listed in Appendix I or provided by your instructor.

PRACTICE 4-43

Butt Joint 4G

Using a properly set up and adjusted FCA welding machine, proper safety protection, E71T-1 and/or E71T-5 electrodes of diameter 0.030 in. and/or 0.035 in. (0.8 mm and/or 0.9 mm), and one or more pieces of

Module 1
Key Indicator 1, 2, 3, 4

Module 2
Key Indicator 1, 2, 3, 4, 7

Module 6
Gas Shielded
Key Indicator 6
Self Shielded
Key Indicator 11
This practice addresses the
"Horizontal" position portion of
the all-position requirement
of 6 and 11.

Module 1
Key Indicator 1, 2, 3, 4

Module 2
Key Indicator 1, 2, 3, 4, 7

Module 6
Gas Shielded
Key Indicator 6
Self Shielded
Key Indicator 11

Module 9
Key Indicator 1, 2
This practice addresses the
"Horizontal" position portion
of the all-position requirement
of 6 and 11.

Module 1
Key Indicator 1, 2, 3, 4

Module 2
Key Indicator 1, 2, 3, 4, 7

Module 6
Gas Shielded
Key Indicator 6
Self Shielded
Key Indicator 11
This practice addresses the
"Overhead" position portion of
the all-position requirement
of 6 and 11.

mild steel sheet, 12 in. (305 mm) long and 16-gauge to 18-gauge thick, you will make a butt weld in the overhead position.

Following the same instructions for the assembly and welding procedure outlined in Practice 4-32, repeat with both classifications of electrodes until defect-free welds can consistently be made. Turn off the welding machine and shielding gas and clean up your work area when you are finished welding.

Complete a copy of the "Student Welding Report" listed in Appendix I or provided by your instructor.

PRACTICE 4-44

Butt Joint 4G 100% to Be Tested

Using a properly set up and adjusted FCA welding machine, proper safety protection, E71T-1 and/or E71T-5 electrodes of diameter 0.030 in. and/or 0.035 in. (0.8 mm and/or 0.9 mm), and one or more pieces of mild steel sheet, 12 in. (305 mm) long and 16-gauge to 18-gauge thick, you will make a butt weld in the overhead position.

Following the same instructions for the assembly and welding procedure outlined in Practice 4-32, repeat the weld until you can use each electrode classification to make welds with 100% penetration that will pass a bend test. Turn off the welding machine and shielding gas and clean up your work area when you are finished welding.

Complete a copy of the "Student Welding Report" listed in Appendix I or provided by your instructor.

Module 1
Key Indicator 1, 2, 3, 4

Module 2
Key Indicator 1, 2, 3, 4, 7

Module 6
Gas Shielded
Key Indicator 6
Self Shielded
Key Indicator 11

Module 9
Key Indicator 1, 2
This practice addresses the "Overhead" position portion of the all-position requirement of 6 and 11.

PRACTICE 4-45

AWS SENSE Entry-Level Welder Workmanship Sample for Flux Cored Arc Welding, Gas-Shielded (FCAW)

Welding Procedure Specification (WPS)

Welding Procedure Specification No.: Practice 4-45. Date:

Module 1
Key Indicator 1, 2, 3, 4

Module 2
Key Indicator 1, 2, 3, 4, 7

Module 6
Key Indicator 7

Title:
Welding FCAW of plate to plate.

Scope:
This procedure is applicable for V-groove, bevel, and fillet welds within the range of 1/8 in. (3.2 mm) through 1-1/2 in. (38 mm).

Welding may be performed in the following positions: all.

Base Metal:
The base metal shall conform to carbon steel M-1, P-1, and S-1, Group 1 or 2.

Backing material specification: none.

Filler Metal:
The filler metal shall conform to AWS specification no. E71T-1 from AWS specification A5.20. This filler metal falls into F-number F-6 and A-number A-1.

Shielding Gas:
The shielding gas, or gases, shall conform to the following compositions and purity:

CO_2 at 30 to 50 cfh or 75% Ar/25% CO_2 at 30 to 50 cfh.

Joint Design and Tolerances:

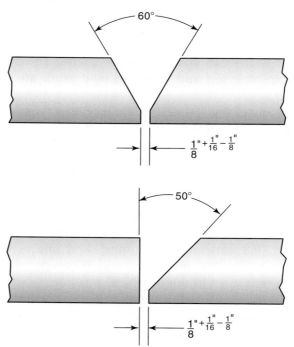

Preparation of Base Metal:

The bevels are to be flame cut on the edges of the plate before the parts are assembled. The beveled surface must be smooth and free of notches. Any roughness or notches deeper than 1/64 in. (0.4 mm) must be ground smooth.

All hydrocarbons and other contaminations, such as cutting fluids, grease, oil, and primers, must be cleaned off all parts and filler metals before welding. This cleaning can be done with any suitable solvents or detergents. The groove face and inside and outside plate surface within 1 in. (25 mm) of the joint must be mechanically cleaned of slag, rust, and mill scale. Cleaning must be done with a wire brush or grinder down to bright metal.

Electrical Characteristics:

The current shall be direct current electrode positive (DCEP). The base metal shall be on the negative side of the line.

Electrode		Welding Power			Shielding Gas		Base Metal	
Type	Size	Amps	Wire-feed Speed, ipm (cm/min)	Volts	Type	Flow	Type	Thickness
E71T-1	0.035 in. (0.9 mm)	130 to 150	288 to 380 (732 to 975)	22 to 25	CO_2 or 75% Ar/ CO_2 25%	30 to 50	Low-carbon steel	1/4 in. to 1/2 in. (6 mm to 13 mm)
E71T-1	0.045 in. (1.2 mm)	150 to 210	200 to 300 (508 to 762)	28 to 29	CO_2 or 75% Ar/ CO_2 25%	30 to 50	Low-carbon steel	1/4 in. to 1/2 in. (6 mm to 13 mm)

Preheat:

The parts must be heated to a temperature higher than 50°F (10°C) before any welding is started.

Backing Gas:

N/A

Safety:

Proper protective clothing and equipment must be used. The area must be free of all hazards that may affect the welder or others in the area. The welding machine, welding leads, work clamp, electrode holder, and other equipment must be in safe working order.

Welding Technique:

Using a 1/2-in. (13-mm) or larger gas nozzle and a distance from contact tube to work of approximately 3/4 in. (19 mm) for all welding, first tack weld the plates together according to Figure 4.58. There should be a root gap of about 1/8 in. (3.2 mm) between the plates with V-grooved or beveled edges. Use an E71T-1 arc welding electrode to make a root pass to fuse the plates together. Clean the slag from the root pass, being sure to remove any trapped slag along the sides of the weld.

Using an E71T-1 arc welding electrode, make a series of stringer or weave filler welds, no thicker than 1/4 in. (6.4 mm), in the groove until the joint is filled. The 1/4-in. (6.4-mm) fillet welds are to be made with one pass.

Interpass Temperature:

The plate should not be heated to a temperature higher than 350°F (175° C) during the welding process. After each weld pass is completed, allow it to cool but never to a temperature below 50°F (10°C). The weldment must not be quenched in water.

Cleaning:

The slag must be cleaned off between passes. The weld beads may be cleaned by a hand wire brush, a hand chipping, a punch and hammer, or a needle-scaler. All weld cleaning must be performed with the test plate in the welding position. A grinder may not be used to remove weld control problems such as undercut, overlap, or trapped slag.

Inspection:

Visually inspect the weld for uniformity and discontinuities. There shall be no cracks, no incomplete fusion, and no overlap. Undercut shall not exceed the lesser of 10% of the base metal thickness or 1/32 in. (0.8 mm). The frequency of porosity shall not exceed one in each 4 in. (100 mm) of weld length, and the maximum diameter shall not exceed 3/32 in. (2.4 mm).

Sketches:

See Figure 4.58.

Complete a copy of the "Student Welding Report" listed in Appendix I or provided by your instructor.

PRACTICE 4-46

AWS SENSE Entry-Level Welder Workmanship Sample for Flux Cored Arc Welding Self-Shielded (FCAW)

Welding Procedure Specification (WPS)

Welding Procedure Specification No.: Practice 4-46 Date:

Title:

Welding FCAW of plate to plate.

Module 1
Key Indicator 1, 2, 3, 4

Module 2
Key Indicator 1, 2, 3, 4, 7

Module 6
Key Indicator 12

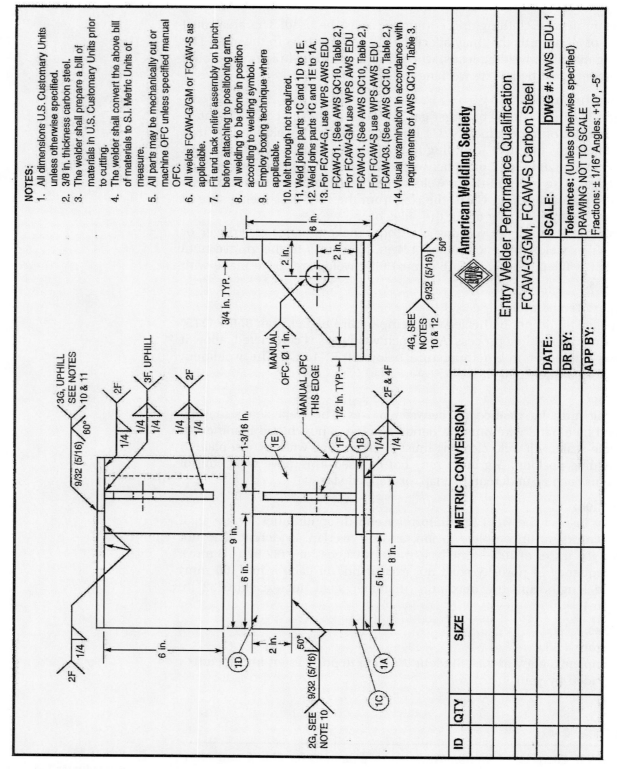

Figure 4.58 FCAW-G workmanship qualification test
Source: Courtesy of the American Welding Society

Scope:

This procedure is applicable for V-groove, bevel, and fillet welds within the range of 1/8 in. (3.2 mm) through 1 1/2 in. (38 mm).

Welding may be performed in the following positions: all.

Base Metal:

The base metal shall conform to carbon steel M-1, P-1, and S-1, Group 1 or 2.

Backing material specification: none.

Filler Metal:

The filler metal shall conform to AWS specification no. 0.035 to 0.0415 dia. E71T-11 from AWS specification A5.20. This filler metal falls into F-number F-6 and A-number A-1.

Shielding Gas:

The shielding gas, or gases, shall conform to the following compositions and purity: none.

Joint Design and Tolerances:

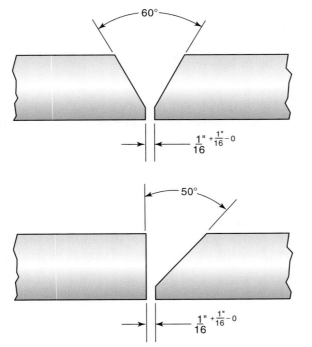

Preparation of Base Metal:

The bevels are to be flame cut on the edges of the plate before the parts are assembled. The beveled surface must be smooth and free of notches. Any roughness or notches deeper than 1/64 in. (0.4 mm) must be ground smooth.

All hydrocarbons and other contaminations, such as cutting fluids, grease, oil, and primers, must be cleaned off all parts and filler metals before welding. This cleaning can be done with any suitable solvents or detergents. The groove face and inside and outside plate surface within 1 in. (25 mm) of the joint must be mechanically cleaned of slag, rust, and mill scale. Cleaning must be done with a wire brush or grinder down to bright metal.

Electrical Characteristics:

The current shall be direct current electrode negative (DCEN). The base metal shall be on the positive side of the line.

| Electrode | | Welding Power | | | Shielding Gas | | Base Metal | |
Type	Size	Amps	Wire-feed Speed, ipm (cm/min)	Volts	Type	Flow	Type	Thickness
E71T-11	0.035 in. (0.9 mm)	130 to 150	150 to 225 (381 to 571)	22 to 25	None	—	Low-carbon steel	1/4 in. to 1/2 in. (6 mm to 13 mm)
E71T-11	0.045 in. (1.2 mm)	150 to 210	105 to 195 (266 to 495)	15 to 18	None	—	Low-carbon steel	1/4 in. to 1/2 in. (6 mm to 13 mm)

Preheat:

The parts must be heated to a temperature higher than 50°F (10°C) before any welding is started.

Backing Gas:

N/A

Safety:

Proper protective clothing and equipment must be used. The area must be free of all hazards that may affect the welder or others in the area. The welding machine, welding leads, work clamp, electrode holder, and other equipment must be in safe working order.

Welding Technique:

Using a 1/2-in. (13-mm) or larger gas nozzle and a distance from contact tube to work of approximately 3/4 in. (19 mm) for all welding, first tack weld the plates together according to Figure 4.59. There should be a root gap of about 1/8 in. (3.2 mm) between the plates with V-grooved or beveled edges. Use an E71T-11 arc welding electrode to make a root pass to fuse the plates together. Clean the slag from the root pass, being sure to remove any trapped slag along the sides of the weld.

Using an E71T-11 arc welding electrode, make a series of stringer or weave filler welds, no thicker than 1/4 in. (6.4 mm), in the groove until the joint is filled. The 1/4-in. (6.4-mm) fillet welds are to be made with one pass.

Interpass Temperature:

The plate should not be heated to a temperature higher than 350°F (175°C) during the welding process. After each weld pass is completed, allow it to cool but never to a temperature below 50°F (10°C). The weldment must not be quenched in water.

Cleaning:

The slag must cleaned off between passes. The weld beads may be cleaned by a hand wire brush, a hand chipping, a punch and hammer, or a needle-scaler. All weld cleaning must be performed with the test plate in the welding position. A grinder may not be used to remove weld control problems such as undercut, overlap, or trapped slag.

Inspection:

Visually inspect the weld for uniformity and discontinuities. There shall be no cracks, no incomplete fusion, and no overlap. Undercut shall not exceed the lesser of 10% of the base metal thickness or 1/32 in. (0.8 mm). The frequency of porosity shall not exceed one in each 4 in. (100 mm) of weld length, and the maximum diameter shall not exceed 3/32 in. (2.4 mm).

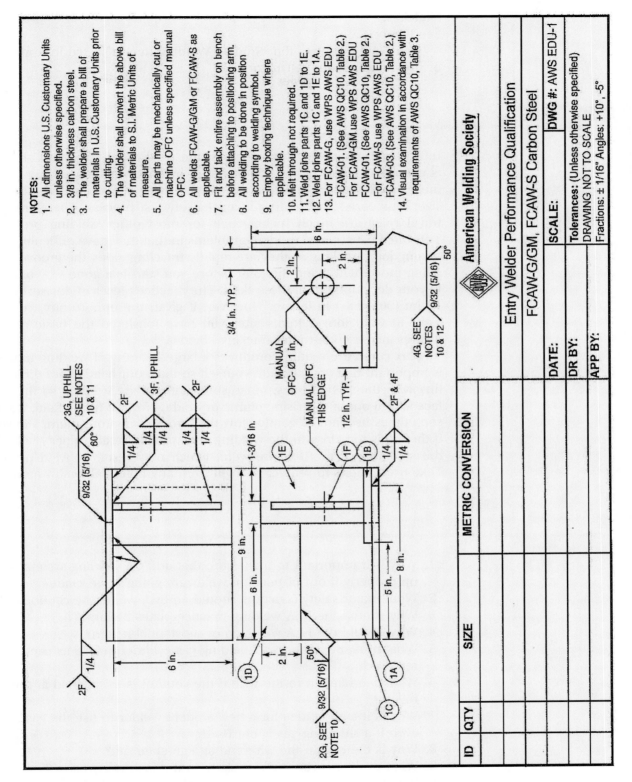

Figure 4.59 FCAW-S workmanship qualification test
Source: Courtesy of the American Welding Society

Sketches:
See Figure 4.59.

Complete a copy of the "Student Welding Report" listed in Appendix I or provided by your instructor.

SUMMARY

In semiautomatic welding processes, the weld travel rate along the joint is controlled more by the process than by your welding technique. You must therefore learn how to travel at the proper rate to maintain the weld size. Flux cored arc welding is a relatively fast process. Therefore, your travel rates are much higher than for most other welding processes. This often causes new welders problems in that they have difficulty maintaining joint tracking as they are rapidly traveling along the groove. Practicing movement along the joint before you start is a good way of aiding in your development of these skills. The practice pieces in this section are 12 in. (305 mm) in length. However, if given the opportunity, you may want to weld longer joints after you have mastered the basic skills to further increase your joint tracking abilities.

Flux cored arc welding produces a large quantity of welding fumes. It is important that you position yourself so that your head is not directly in line with the rising fumes. Make sure that you are welding so that your face is well out of this rising plume of welding fumes. In the field, welders sometimes use fans to gently blow the fumes away from them. However, if the fan is too close to the welding zone, excessive air velocity will blow the shielding away from the weld, which may result in weld porosity. Take precautions to protect yourself from any potential health hazards.

REVIEW

1. Why is it important to make sure that an FCA welding system is set up properly if out-of-position welds are going to be made?
2. What major safety concerns should an FCA welder be cautious of?
3. Why should the FCA welding practice plates be large?
4. Why should the FCA welds be of substantial length?
5. What must be done to a shielding gas cylinder before its cap is removed?
6. What can happen to the wire if the conduit is misaligned at the feed rollers?
7. Why is it a good idea for a new student welder to use the gas nozzle even if a shielding gas is not used?
8. Why is the curl in the wire end straightened out?
9. What problems can high pressure in the feed roller cause?
10. Referring to Table 4.1, answer the following:
    a. What would the range of the feed speed be for an amperage of 150 at 25 volts for an E70T-1 0.035-in. (0.9-mm) electrode?
    b. What would the approximate amperage be for an E71T-11 0.045-in. (1.2-mm) electrode if it is being fed at a rate of 200 in. per minute (508 cm per minute)?

11. What are the disadvantages of beveling plates for welding?
12. FCA welds with 100% joint penetration can be made in plates up to what thickness?
13. What is the smallest V-groove angle that can be welded using the FCA welding process?
14. What is the purpose of the root pass?
15. Why is the FCA welding process not used for open-root critical welds?
16. Why should convex weld faces be avoided?
17. What bead pattern is best for overhead welds?
18. Why is the appearance of the cover pass so important?
19. What is the visual inspection standard's limitations of acceptance?
20. Why is the backing strip 2 in. (50 mm) longer than the test plate?
21. Why should there be a space between the plates when making a fillet weld on thick plates of a tee joint?
22. What would a small notch at the root weld's leading edge in a fillet weld mean?
23. What changes should be made in the setup for making a vertical up weld?
24. How can a cold weld be corrected?
25. What problem must be overcome if the amperage and voltage are lowered to make a vertical weld?
26. How can higher welding speeds help control distortion?
27. How can spatter buildup on the welding gun be controlled in the overhead position?

APPENDIX I—STUDENT WELDING REPORT

I. STUDENT WELDING REPORT

Student Name: _____ Date: _____

Instructor: _____ Class: _____

Experiment or Practice #: _____ Process: _____

Briefly describe task: _____

_____

_____

_____

INSPECTION REPORT			
Inspection	Pass/Fail	Inspector's Name	Date
Safety:			
Equip. Setup:			
Equip. Operation:			
Welding	Pass/Fail	Inspector's Name	Date
Accuracy:			
Appearance:			
Overall Rating:			

Comments:

Student Grade: _____ Instructor Initials: _____ Date: _____

Glossary

air-cooled Welding guns or torches that are cooled by ambient air or shielding gas passing through the system.

enfriado por aire Pistolas o antorchas de soldadura que son enfriadas por aire ambiental o protección de gas que pasa a través del sistema.

amperage range The lower and upper limits of welding power, in amperage, that can be produced by a welding machine or used with an electrode or by a process.

rango de amperaje Los límites máximos y mínimos de poder de soldadura (en amperaje) que puede tener una máquina para soldar o que pueden usarse con un electrodo o a través de un proceso.

bird-nesting A condition where GMAW or FCAW filler wire becomes tangled in the wire feeding unit.

enredado de alambre Una condición GMAW o FCAW donde el relleno de alambre se enreda en la unidad alimentadora de alambre.

cast The natural curve in the electrode wire for gas metal arc welding as it is removed from the spool; cast is measured by the diameter of the circle that the wire makes when it is placed on a flat surface without any restraint.

distancia La curva natural en el alambre electrodo para soldadura de arco metálico para gas cuando se aparta del carrete; la distancia es medida en el círculo que hace el alambre cuando es puesto en una superficie plana sin restricción.

coils Large volume GMAW/FCAW electrode wire holder that is larger than a spool and smaller than a drum.

bobinas Sujetador de grandes volúmenes de alambre electrodo GMAW/FCAW que es más grande que un carrete y más pequeño que un tambor.

conduit liner A flexible steel tube that guides the welding wire from the feed rollers through the welding lead to the gun used for GMAW and FCAW welding. The steel conduit liner may have a nylon or Teflon® inner surface for use with soft metals such as aluminum.

revestimiento de conducto Un tubo flexible de acero que guía el alambre para soldar desde los rodillos de alimentación, a través de los cables para soldar, hasta la pistola, usado en soldaduras de tipo GMAW y FCAW. El revestimiento del conducto de acero puede tener una superficie interior de Teflon® o nylon para su uso con metales blandos como el aluminio.

contact tube A device that transfers current to a continuous electrode.

tubo de contacto Un aparato que traslada corriente continua a un electrodo.

critical weld A weld so important to the soundness of the weldment that its failure could result in the loss or destruction of the weldment and injury or death.

soldadura crítica Una soldadura tan importante para la calidad del conjunto de partes soldadas, que su fracaso podría ocasionar la pérdida o destrucción de dicho conjunto, así como también lesiones o muerte.

deoxidizers Chemical additions made to fluxes or solid filler wires that are intended to reduce discontinuities such as porosity and cracking on base metals with mill scale or limited pre-cleaning.

deoxidantes Agregado de productos químicos a alambres llenados de fluídos o sólidos con el propósito de disminuir pérdidas como porosidad y grietas en metales base con óxido de hierro con o limpieza previa limitada.

dual shield Welding industry term synonymous with FCAW electrodes that require external gas shielding.

doble proteccion Los términos sinónimos de la industria de soldadura con electrodos FCAW que requieren protección de gas externa.

electrode extension (GMAW, FCAW, SAW) The length of unmelted electrode extending beyond the end of the contact tube during welding.

extensión del electrodo (GMAW, FCAW, SAW) La distancia de extensión del electrodo que no está derretido más allá de la punta del tubo de contacto durante la soldadura.

feed rollers A set of two or four individual rollers which, when pressed tightly against the filler wire and powered up, feed the wire through the conduit liner to the gun for GMAW and FCAW welding.

rodillos de alimentación Un conjunto de dos o cuatro rodillos individuales que al ser presionados fuertemente contra el alambre de relleno y ser accionados alimentan al alambre a través del revestimiento de canal hasta la pistola, en soldaduras tipo GMAW y FCAW.

flow rate The rate at which a given volume of shielding gas is delivered to the weld zone. The units used for welding are cubic feet, inches, meters, and centimeters.

caudal Velocidad a la cual llega un determinado volumen de gas protector a la zona de soldadura. Las unidades usadas para la soldadura son pies cúbicos, pulgadas, metros, y centímetros.

flux cored arc welding (FCAW) An arc welding process that uses an arc between a continuous filler metal electrode and the weld pool. The process is used with shielding gas from a flux contained within the tubular electrode, with or without additional shielding from an externally supplied gas, and without the application of pressure.

soldadura de arco con núcleo de fundente   Un proceso de soldadura de arco que usa un arco entre medio de un electrodo de metal rellenado continuo y el charco de la soldadura. El proceso es usado con gas de protección del flujo contenido dentro del electrodo tubular, y sin usarse protección adicional de abastecimiento de gas externo, y sin aplicarse presión.

globular transfer   The transfer of molten metal in large drops from a consumable electrode across the arc.

traslado globular   El traslado del metal derretido en gotas grandes de un electrodo consumible a través del arco.

lap joint   A joint between two overlapping members.

junta de solape   Una junta entre dos miembros traslapadas.

lime-based flux   Basic type flux system associated with T-5 FCAW electrodes and suitable in the flat and horizontal fillet positions.

fundente a base de óxido de calcio   Sistema fundente del tipo básico relacionado con electrodes T-5 FCAW y apropiado para las posiciones de filete plana y horizontal.

pinch effect   The phenomena in GMAW short circuiting transfer that takes place when the molten droplet of filler metal detaches from the electrode wire.

efecto de compresión   El fenómeno en transferencia de corto circuito en GMAW tiene lugar cuando el fundido en gotas de metal o relleno se desprende desde el alambre electrodo.

pulsed-arc metal transfer   In gas metal arc welding, pulsing the current from a level below the transition current to a level above the transition current to achieve a controlled spray transfer at lower average currents; spray transfer occurs at the higher current level.

transferir el metal por arco pulsado   En la soldadura de arco metálico con gas, se pulsa la corriente de un nivel más alto de la corriente de transición para lograr un traslado de rocío controlado a una corriente media baja; el traslado del rocío ocurre al nivel más alto de la corriente.

root face   The portion of the groove face adjacent to the root of the joint.

cara de raíz   La porción de la cara de la ranura adyacente a la raíz de la junta.

rutile-based flux   Acidic T1 type FCAW flux component capable of all position operations.

fundente a base de rutilo   Fundente ácido tipo T1 FCAW (por sus siglas en inglés soldadura por arco con núcleo fundente) eficiente en todo lugar de operaciones.

self-shielding   FCAW electrodes that do not require external gas shielding.

auto-protección   Los electrodos FCAW que no requieren gas de protección externa.

short-circuiting transfer   (arc welding) Metal transfer in which molten metal from a consumable electrode is deposited during repeated short circuits.

transferir por corto circuito   (soldadura de arco) Transferir metal el cual el metal derretido del electrodo consumible es depositado durante repetidos cortos circuitos.

slag   A nonmetallic product resulting from the mutual dissolution of flux and nonmetallic impurities in some welding and brazing processes.

escoria   Un producto que no es metálico resultando de una disolución mutual del flujo y las impurezas no metálicas en unos procesos de soldadura y soldadura fuerte.

slope   For gas metal arc welding, the volt-ampere curve of the power supply indicates that there is a slight decrease in voltage as the amperage increases; the rate of voltage decrease in the slope.

pendiente   Para soldadura de arco de metal con gas, la curva voltio-amperio de la fuente de poder indica que si hay un ligero decremento en voltaje cuando los amerios aumentan; la proporción del voltaje decrementa en el pendiente.

smoke extraction nozzles   Special FCAW guns that have built-in fume extraction capabilities suited for indoor use of FCAW self shielded electrodes.

boquillas de extracción de humo   Las pistolas especiales FCAW fabricadas con capacidades para extracción de humo adaptadas para uso interno de electrodos FCAW con protección propia.

spray transfer   (arc welding) Metal transfer in which molten metal from a consumable electrode is propelled axially across the arc in small droplets.

traslado rociado   (soldadura de arco) Transferir el metal el cual el metal derretido de un electrodo consumible es propelado axialmente a traves del arco en gotitas pequeñas.

spool drag   The amount of tension set on a spool of GMAW/FCAW electrode wire to keep slack from developing between the electrode wire spool and the wire feeder.

arrastre de carrete   La cantidad de tensión ejercida en un carrete de alambre electrodo GMAW/FCAW para mantenerlo flojo desde el desarrollo entre el carrete de alambre electrodo y el alimentador de alambre.

spools   GMAW/FCAW electrode wire holders.

carretes   Sujetadores de alambre electrodo GMAW (por sus siglas en inglés soldadura por arco metálico con gas)/FCAW.

synergic system   Pulsed-arc metal transfer system in which the power supply and wire-feed settings are made by adjusting a single knob.

sistema sinérgico   Sistema de transferir metal por arco pulsado en cual la fuente de poder y los ajustes del alimentador de alambre son hechos por el ajuste de un botón solamente.

stringer bead   A type of weld bead made without appreciable weaving motion.

cordón encordador Un tipo de cordón de soldadura sin movimiento del tejido apreciable.

tee joint A joint between two members located approximately at right angles to each other in the form of a *T*.

junta en T Una junta en medio de dos miembros que están localizados aproximadamente a ángulos rectos de uno al otro en la forma de *T*.

transition current In gas metal arc welding, current above a critical level to permit spray transfer; the rate at which drops are transferred changes in relationship to the current. Transition current depends upon the alloy bearing welded and is proportional to the wire diameter.

corriente de transición En soldadura de arco y metal con gas, corriente arriba de un nivel crítico para permitir el traslado del rociado; la proporción en la cual las gotas son transferidas cambia en relación a la corriente. La corriente de transición depende del aleado que se está soldando y es proporcional al diámetro del alambre.

voltage range The lower and upper limits of welding power, in volts, that can be produced by a welding machine or used with an electrode or by a process.

rango de voltaje Los límites máximos y mínimos de poder de soldadura (en voltios) que puede tener una máquina para soldar o que pueden usarse con un electrodo o a través de un proceso.

water-cooled Welding guns or torches that are cooled by an external fluid passing through the system.

enfriado por agua Las pistolas o antorchas de soldadura que son enfriadas por un fluido externo que pasa a través del sistema.

weave bead A type of weld bead made with transverse oscillation.

cordón tejido Un tipo de cordón de soldadura hecha con oscilación transversa.

welding helmet A device designed to be worn on the head to protect eyes, face, and neck from arc radiation, radiated heat, spatter, or other harmful matter expelled during arc welding, arc cutting, and thermal spraying.

casco Un aparato diseñado para usarse sobre la cabeza para proteger ojos, cara y cuello de radiación del arco, calor radiado, salpicadura, u otra materia dañosa despedida durante la soldadura de arco, corte por arco, y rociado termal.

wire-feed speed The rate at which wire is consumed in arc cutting, thermal spraying, or welding.

velocidad de alimentador de alambre La velocidad que el alambre es consumido en cortes de arco, rociado termal o soldadura.

Index

Italic page numbers indicate material in tables or figures.